Michael Adler

Monte Carlo simulations of the Ising model

Anchor Academic
Publishing

Adler, Michael: Monte Carlo simulations of the Ising model, Hamburg, Anchor Academic Publishing 2016

Buch-ISBN: 978-3-95489-879-4
PDF-eBook-ISBN: 978-3-95489-466-6
Druck/Herstellung: Anchor Academic Publishing, Hamburg, 2016

Bibliografische Information der Deutschen Nationalbibliothek:
Die Deutsche Nationalbibliothek verzeichnet diese Publikation in der Deutschen Nationalbibliografie; detaillierte bibliografische Daten sind im Internet über http://dnb.d-nb.de abrufbar.

Bibliographical Information of the German National Library:
The German National Library lists this publication in the German National Bibliography. Detailed bibliographic data can be found at: http://dnb.d-nb.de

All rights reserved. This publication may not be reproduced, stored in a retrieval system or transmitted, in any form or by any means, electronic, mechanical, photocopying, recording or otherwise, without the prior permission of the publishers.

Das Werk einschließlich aller seiner Teile ist urheberrechtlich geschützt. Jede Verwertung außerhalb der Grenzen des Urheberrechtsgesetzes ist ohne Zustimmung des Verlages unzulässig und strafbar. Dies gilt insbesondere für Vervielfältigungen, Übersetzungen, Mikroverfilmungen und die Einspeicherung und Bearbeitung in elektronischen Systemen.

Die Wiedergabe von Gebrauchsnamen, Handelsnamen, Warenbezeichnungen usw. in diesem Werk berechtigt auch ohne besondere Kennzeichnung nicht zu der Annahme, dass solche Namen im Sinne der Warenzeichen- und Markenschutz-Gesetzgebung als frei zu betrachten wären und daher von jedermann benutzt werden dürften.

Die Informationen in diesem Werk wurden mit Sorgfalt erarbeitet. Dennoch können Fehler nicht vollständig ausgeschlossen werden und die Diplomica Verlag GmbH, die Autoren oder Übersetzer übernehmen keine juristische Verantwortung oder irgendeine Haftung für evtl. verbliebene fehlerhafte Angaben und deren Folgen.

Alle Rechte vorbehalten

© Anchor Academic Publishing, Imprint der Diplomica Verlag GmbH
Hermannstal 119k, 22119 Hamburg
http://www.diplomica-verlag.de, Hamburg 2016
Printed in Germany

Monte Carlo Methods applied to
the Ising model

Michael Adler

The thermodynamic observables of the classical one– and two–dimensional ferromagnetic and antiferromagnetic Ising models on a square lattice are simulated, especially at the phase transitions (if applicable) using the classical Monte Carlo algorithm of Metropolis. Finite size effects and the influence of an external magnetic field are described. The critical temperature of the 2d ferromagnetic Ising model is obtained using finite size scaling.

"My parents"

Contents

1. **Statistical mechanics – a short review** 3
   1.1. Description of state of a system 4
      1.1.1. Phase space 4
      1.1.2. Statistical ensembles 5
   1.2. Transition between different states 7
   1.3. Thermodynamic observables 8
   1.4. Phase transitions 10
   1.5. n-vector models 12
      1.5.1. Ising model 13

2. **Monte Carlo simulation** 17
   2.1. Fundamentals 17
      2.1.1. Crude Monte Carlo 18
      2.1.2. Markov Chain Monte Carlo 19
      2.1.3. Markov chains 21
   2.2. Monte Carlo simulations in statistical mechanics 22
      2.2.1. Ergodicity and detailed balance 22
      2.2.2. Acceptance ratios 23
      2.2.3. Metropolis algorithm 23
      2.2.4. Initialization bias 24
      2.2.5. Autocorrelation in equilibrium 25
   2.3. Analyzing the data – T_c and more 26
      2.3.1. Finite size scaling and the Binder ratio 26

3. **Selected results** 29
   3.1. One–dimensional Ising model 29
   3.2. Two–dimensional Ising models 33
      3.2.1. The ferromagnetic Ising model 33
      3.2.2. Antiferromagnetic Ising model 39

Contents

3.3.	The phase transition	44
3.4.	The Critical temperature and critical exponents	45
	3.4.1. Binder cumulant	45
	3.4.2. Critical exponents	46

A. Source code of MC integrators   **49**
- A.1. Crude Monte Carlo 49
- A.2. Markov chain MC integrator 50

B. Implementation of the Ising models   **55**
- B.1. Skeleton of the code 55
- B.2. Source codes of the Ising models 57
  - B.2.1. One dimensional Ising model 57
- B.3. Two dimensional Ising model 70

Bibliography   **91**

List of Figures   **93**

Chapter 1.

Statistical mechanics – a short review

Before presenting the Ising model we will refresh the basic concepts of statistical mechanis. For more information on this important topic the reader is referred to (9, 22)

In thermodynamics we are interested in the properties of systems consisting of a large number of particles. However the sheer number of particles makes it impossible to write down and solve about 10^{23} coupled equations of motion. We would also need the same number of initial conditions. Obviously it is also impossible to get experimentally all the initial conditions required. But we are not interested in the microscopic properties (EOMs of the particles of our system) but rather in *macroscopic observables* like the energy, temperature, heat capacity etc. Hence it is sufficient to have only information about the *statistical* properties of our system. Henceforth we will use the terms *microscopic* and *macroscopic* systems. A macroscopis system consisting of N[1] particles fulfills:

$$\frac{1}{\sqrt{N}} \ll 1$$

Any system that does not fulfill this conditions is microscopic. Applications of statistical methods to such a system would entail unreasonable errors.

---

[1] N is typically $6 \cdot 10^{23}$ or more.

1.1. Description of state of a system

1.1.1. Phase space

As stated above we are interested in the statistical properties of our system. Thus it is sufficient to have knowledge about the average motion of particles in our system. Instead of considering the motion of only one system we can consider an *ensemble* consisting of copies of the system we are interested in. The copies only differ in the initial values of their motion.

As an alternative we could take a sufficiently long time and determine the average of the interesting quantity. We will however, follow the way proposed by the American physicist GIBBS. Instead of the time-average he took the average over many similar systems all having the same macroscopic properties like internal energy, pressure, ... This means that all elements of the ensemble are *in states accessible to the system*. The fact that we can decide which average (time or ensemble average) we'll consider is far from trivial – we'll later give the details.

Classically the state of a system is fully described by the generalized coordinates and momenta $\vec{q}_i$ and $\vec{p}_i$ of every particle i; with $\vec{p} = (p_x, p_y, p_z)$ and analogously for q. The description is only complete if we know the equations of motion for the particles. In the six-dimensional *phase space* $\Gamma = \text{span}\{\vec{p}, \vec{q}\}$, i.e. the space spun by the vectors of momentum and space every state of our macrosystem is represented by a point in phase space:

$$\vec{\pi} = (\pi_1, \pi_2, \cdots, \pi_{2s}) = (\vec{q}, \vec{p})$$

Due to errors in measurement of p and q the system is specified by a cell of area h_0 in phase space. Thus:

$$\overline{(\Delta p)^2} \cdot \overline{(\Delta q)^2} \geq h_0$$

Quantum mechanics sets the lower boundary for h_0: $h_0 = \hbar/2$. In quantum mechanics the state of the system is specified by a wave-function at time t:

$$\Psi(\vec{q}, \vec{s}, t) \quad \vec{s} \text{ internal degrees of freedom}$$

Statistical mechanics – a short review

$\vec{s}$ can be for instance the spin of the particle. Now we introduce the propability density ρ, i. e. the propabiliy per volume in phase space of finding the system at time t_0 in state x:

$$p_x = \rho_x(t_0) \cdot (\Delta p)^s (\Delta q)^s \qquad (1.1)$$

In the classical regime ρ is (9):

$$\rho_x \stackrel{\text{def}}{=} \frac{1}{I} \sum_i \delta(x - x^{(i)}) \delta(p - p^{(i)}) \qquad (1.2)$$

1.1.2. Statistical ensembles

In general we specify certain macroscopic properties like the total internale energy E, the temperature T, the pressure p,... of our system. Of course the values of the observables are subject to fluctuations due to the finite precision of all possible measurements. In general, however the obversables are time-dependent. If our system is in equilibrium its observables are no longer time-dependent. The time until the system has reached equilibrium is called *equilibration time* τ_{eq}. The fundamental postulat of statistical mechanics, the *principle of equal a-priori propabilities* states that:

The system[2] is equally likely to be found in any of its accessible states.

This holds only for macroscopic systems in equilibrium. This postulate cannot be proven rigorously.

There are different possible restraints for the system. The simplest is to specify that the internal energy E_n has to be whithin the interval: $(E - \Delta E, E]$. If we require an isolated system, i.e. $N = $ const. and $V = $ const. we'll get the **microcanonical ensemble:** Ω denotes the number of accessible microstates. The chance of finding a system chosen by random from the ensemble in a given microstate is $\frac{1}{\Omega}$. Thus the propability density $\rho(E_n)$ that depends on the internal Energy E_n of the macrosystem is:

$$\rho(E_n) = \begin{cases} \frac{1}{\Omega} & \text{if: } E - \Delta E \leq E_n \leq E, \\ 0 & \text{else.} \end{cases} \qquad (1.3)$$

---

[2] We refer here to an isolated system.

The number of possible states at constant E, N and V is called microcanonical partitition function Z_m:

$$Z_m = \sum_{E_n(N,V) < E} \qquad (1.4)$$

E_n is the energy of the n-th state at given E, N and V. In a later section we'll see that the knowledge of an analytical partition function is tantamount to knowledge of the thermodynamic observables.

In experiment however the situation of an isolated system is rather an exception. If we allow exchange of thermal energy with a heat reservoir we'll get a more feasible ensemble, the **canonical ensemble:** The number of particles in the heat bath N' has to be much greater than the number of particles N in the system of interest. It should be noted that volume and temperature of our system of interest as well as of the heat reservoir (obviously) are constant. Since we are considering a system that exchanges heat with a reservoir we'll have a more complicated partition function. The propability density is:

$$\rho_n = \frac{1}{Z} e^{-\beta E_n} \qquad (1.5)$$

With the partition function:

$$Z_c = \sum_n e^{-\beta E_n} \qquad (1.6)$$

We call $\exp(-\beta E_n)$ BOLTZMANN factor and $\beta = 1/kT$ reduced temperature. In terms of the Hamilton operator we may write: $\rho_n = 1/Z \exp(-\beta H)$ and $Z_c = \text{tr}(-\beta H)$.

Going one step further loosening the restraints on the system of interest we allow exchange of particles with a heat reservoir. Only the total energy and the number of particles of both systems together are conserved. We find:

$$\rho_n(E_n, N_n) = \frac{1}{Z} e^{-\beta(E_n - \mu N_n)} \qquad (1.7)$$

Statistical mechanics – a short review

We expect a more general partition function containing the canonical partition function:

$$Z = \sum_n e^{-\beta(E_n - \mu_n N)} \tag{1.8}$$

The change in energy due to adding a particle to the system if entropy and volume are held fixed is given by the chemical potential μ. Unless indicated differently we'll use the canonical ensemble. It has to be stressed, however that they are equivalent in the thermodynamic limit; i.e. if $N \to \infty$.

1.2. Transition between different states

Now we'll introduce the important concept of the *master equation*. Suppose our system is in a state i at time t. We denote the probability of a transition to another state j Δt later by T_{ij}. We'll assume that T_{ij} is time-independent. The time-dependent probability of finding the system in a state i is $w_i(t)$. Since every systems is in a certain state:

$$\sum_i w_i(t) = 1 \tag{1.9}$$

Now we can write down the master equation:

$$\frac{dw_i}{dt} = \sum_j (T_{ji} w_i - T_{ij} w_j) \tag{1.10}$$

The master equation describes the time-dependent change of the propability for finding our system in a state i. Thus the first term of the right-hand side of the master equation describes the rate of transitions from other states to state i. The transition from state j to other states is represented by the second term on the right-hand side. In equilibrium the left-hand side of the master equation vanishes and we get the equilibrium occupation propability p_i

$$p_i = \lim_{t \to \infty} w_i(t) \tag{1.11}$$

We can state now the equation for the average of an observable $\mathcal{O}$:

$$\langle \mathcal{O} \rangle = \sum_i \mathcal{O}_i p_i \tag{1.12}$$

$\mathcal{O}_i$ is the value of $\mathcal{O}$ if the system is in state i. The propability for this is p_i. A detailed exposition of the important topic of master equations is given in (9).

1.3. Thermodynamic observables

Once we hav obtained the partition function it is straightforward to calculate thermodynamic observables. They describe the macroscopic properties of our system and we are primarily interested in them. In most cases we won't be lucky having knowledge about the partition function. Ways to calculate thermodynamic observables in this instance will be explained in later chapters. As far as macrosystem are concerened all three mentioned partition functions may be used.

The *expectation value* of an observable $\mathcal{O}$ is defined as:

$$\langle \mathcal{O} \rangle = \frac{1}{Z} \sum_i \exp(-\beta E_i) \mathcal{O}_i \tag{1.13}$$

Internal energy: This is the most important observable it is denoted by U. However since we are interested in $\langle E \rangle$ we'll drop the brackets and use E meaning the average energy in later section dealing only with our results and therefore only with averages.

$$U = \langle E \rangle = \frac{1}{Z} \sum_n E_n \exp(-\beta E_n) = -\frac{1}{Z} \frac{\partial Z}{\partial \beta} = -\frac{\partial \ln Z}{\partial \beta} \tag{1.14}$$

Now we may introduce the **heat capacity C**:

$$C = \frac{\partial U}{\partial T} = -k\beta^2 \frac{\partial U}{\partial \beta} = k\beta^2 \frac{\partial^2 \log Z}{\partial \beta^2} \tag{1.15}$$

C is a measure for the amount of heat ΔQ required to change a body's temperature by a given temperature ΔT. Since we can write

$$\langle E^2 \rangle = \frac{1}{Z} \sum_n E_n^2 \exp(-\beta E_n) = \frac{1}{Z} \frac{\partial^2 Z}{\partial \beta^2} \tag{1.16}$$

Statistical mechanics – a short review

and

$$\langle E^2 \rangle - \langle E \rangle^2 = \frac{1}{Z}\frac{\partial^2 \log Z}{\partial \beta^2} \qquad (1.17)$$

it follows

$$C = k\beta^2 \langle E^2 \rangle - \langle E \rangle^2. \qquad (1.18)$$

The next observable of interest is the **total magnetization** M:

$$M = \frac{\mathrm{d}\mu}{\mathrm{d}V}. \qquad (1.19)$$

If we use the equation for the potential magnetic energy

$$E = -\vec{\mu} \cdot \vec{B}, \qquad (1.20)$$

it follows

$$M_n = -\frac{\partial E_n}{\partial B_i}.^3 \qquad (1.21)$$

We can calculate the average magnetization directly by summing over the spins s_i:

$$\langle M \rangle = \langle \sum_i s_i \rangle \qquad (1.22)$$

The average magnetization per spin is just $\frac{1}{N}\langle M \rangle$ with N being the number of spins. It characterizes the strength of magnetism in a certain substance. Another very important observable that is connected with M is the **magnetic suscebptibiliy** χ_m[4]:

$$\chi = \frac{\partial M}{\partial B} \qquad (1.23)$$

It measures the change of magnetization due to an external magnetic field. In the most general case it cannot be described by a scalar but a tensor has to be used due to anistropy of the material. **Entropy** S:

$$S = k_B \ln \Omega \qquad (1.24)$$

---

[4]We'll use χ without the subscricpt m since we are only dealing with magnetic susceptibilities. In the general case, however, a susceptibity $\partial \langle X \rangle / \partial Y$ describes the strength of response of X to a change in Y.

Ω is the number of accessible microstates. So the entropy is a measure for the number of accessible microstates of a given macroscopic system in a specified state. Because of the postualate of equal a priori propabilities the number of accessible states is maximized for an isolated system in equilibrium. Thus S is maximal in equilibrium for an isolated system (or a sufficiently large system). Now we can define the **free energy F**:[5]

$$F = U - TS = -k_B T \log Z \qquad (1.25)$$

It describes the maximal work that may be obtained from an isolated system with $V = \text{const.}$ and $T = \text{const.}$

1.4. Phase transitions

[6] The transition between the solid, liquid and gaseous phase, the transition between normalconductors and superconductors or the transition between ferromagnetism and paramagnetism are well–known examples of *phase transitions*; i. e. transition from a normal phase to an ordered phase. They can be classified according to their order of transition. First order transitions involve *latent heat*: The system interacts with its environment and thereby absorbs or releases a fixed amount of energy without change of its temperature. First order phase transitions are associated with mixed-phase regimes, i. e. only one part of the substance has undergone phase transition. An example is boiling water. Second order phase transitions do not entail latent heat which is why they are also referred to as *continuous phase transitions*. It is this type of phase transitions we are interested in. The free energy has a singularity at the transition; this can be seen from the power-law behavior of observables calculated from F. Another feature is the diverging *correlation length* ζ. It is a measure for the order and correlation in a system. The correlation function $G(x)$ itself measures the order in a system, i. e. it describes the way microscopic variables are correlated at different places, e. g. spins:

$$G(x) = \langle s(x)s(x+x')\rangle - \langle s(x)\rangle^2 \qquad (1.26)$$

---

[5]Note that the IUPAP recommends the name HELMHOLTZ *energy* connected with the letter A instead of F.
[6]For more details see for instance (*18*).

Statistical mechanics – a short review

For $T \sim T_C$ we find typically[7]

$$G \sim \frac{1}{r} \cdot g(r/\zeta) \text{ with } g(r/\zeta) \sim \exp(-r/\zeta) \text{ for } r \to \infty. \qquad (1.27)$$

To measure the degree of order in a system e. g. magnetic order, the order parameter λ is introduced.

$$\lambda = \begin{cases} 0 & T < T_C \\ \lambda' & T \leq T_C \end{cases} \qquad (1.28)$$

In magnetic system the order parameter is the (total) magnetization of the system. In the ferromagnetic regime it has a value $\neq 0$. Below the *Curie temperature* which is the critical temperature of ferromagnetic systems, ferromagnetism occurs because permanent magnetic moments line up parallel. At higher temperatures thermodynamic fluctuations destroy the magnetic order and $M = 0$. Close to T_C observables $\mathcal{O}$ can be described by a power law in first order since higher order contributions are negligible.

$$\mathcal{O} \approx a\epsilon^{-\eta} \text{ with } \epsilon = \left| \frac{T}{T_C} - 1 \right| \qquad (1.29)$$

ϵ is the reduced distance from the critical temperature. Near phase transitions the following relations hold:

1. $C = C_0 \epsilon^{-\alpha}$,
2. $m = m_0 \epsilon^{\beta}$,
3. $\chi = \chi_0 \epsilon^{-\gamma}$,
4. $\zeta = \zeta_0 \epsilon^{-\nu}$.

The fact that the critical exponents are independent of whether $T < T_C$ or $T > T_C$ is justified by the *scaling relations*. Using these relations it can be shown that there are in fact only two independent exponents for the two-dimensional Ising model introduced later.

Another interesting feature of phase transitions is *universality* ([18]): The critical exponents we introduced above depend only on three parameters of a given system:

---

[7] Note that there are *two* correlation length and also two functions $g(r)$ depending on whether $T > T_C$ or $T < T_C$. This general case is only realized in anisotropic systems which is why will ignore it and assume just *one* value.

1. Dimension d of system

2. Range of interaction

3. Dimension of spin

The second and third parameter require explanation. The range of interaction follows

$$r^{-(d+2+x)}.\qquad(1.30)$$

x is constant[8] The term spin dimension will be explained in the next section. Though there are strong arguments for universality of critical exponents it has not (yet?) been proven rigorously.

1.5. n-vector models

The Ising model that the thesis on hand will examine belongs to the class of n-vector models with $n = 1$. They are all defined on a lattice and may under certain circumstances be used to describe phenomena like ferromagnetism. In the following Hamiltonian the indices i and j are used to refer to different spins on the lattice. In the most general Hamiltonian we assume a pair interaction between all spins. The Hamiltonian for the general spin model is:[9]

$$\mathcal{H} = -\sum_{i,j} J_{i,j} \vec{s}_i \vec{s}_j - \vec{H} \sum_i \vec{s}_i \qquad(1.31)$$

with the coupling constant J_{ij}

$$J_{ij} \begin{cases} > 0 & \text{ferromagnetic} \\ < 0 & \text{antiferromagnetic} \\ = 0 & \text{non-interacting} \end{cases} \qquad(1.32)$$

We'll deal with ferromagnetic systems, i.e. $J_{ij} > 0$, furthermore we assume that the coupling constant does not depend on the position on the lattice. Later we'll examine

---
[8]$x > 0$ is referred to as long-ranged interaction; $x < d/2 - 2 < 0$ is called short-ranged-interaction.
[9]See (18).

also antiferromagnetic systems. For the sake of notation we set $J_{ij} \equiv J \equiv 1$. Additionally we restrict the interaction of spins to next neighbors. This assumption is essential if we want to solve the model. This is indicated using $\langle i,j \rangle$ or sometimes (i,j) in the sum:

$$\mathcal{H} = -\sum_{\langle i,j \rangle} \vec{s}_i \vec{s}_j - \vec{H}_0 \sum_i \vec{s}_i \quad (1.33)$$

Depending on the dimension n^{10} of the spin $\vec{s}$ we'll get different n-vector models:

$$n = \begin{cases} 1 & \text{Ising model} \\ 2 & \text{XY model} \\ 3 & \text{Heisenberg model} \end{cases} \quad (1.34)$$

The first and the second model are exactly solved for next neighbor interaction and for $B_0 = 0$. The first has also been solved for $B_0 \neq 0$. For the Heisenberg model so far no analytical solution is known. The Heisenberg model, however, may be tackled numerically with very high precision.

1.5.1. Ising model

The Ising model was originally invented by WILHELM LENZ (14). ERNST ISING solved this model in one dimension in his PhD thesis (11) thereby making it known as Ising model. The one dimensional Ising model has a phase transition from paramagnetism to ferromagnetism at $T = 0$. LARS ONSAGER solved the two dimensional Ising model finding a phase transition at $T \neq 0$ (19).

The Hamiltonian in *one dimension* on a finite lattice is

$$\mathcal{H} = -\sum_{i=1}^{N-1} s_i s_{i+1} - \sum_i s_i \quad (1.35)$$

if we assume periodic boundary conditions, i.e. $s_1 = s_N$. Thereby equivalence of all sites is ensured and the system is translationally invariant (4). s is one–dimensional thus $s = \pm 1$ and it is straightforward to analyze the model. The general relation for M is

---

[10] This is the n of the n-Vector model.

(*18*):

$$M(T, B_0) = N\mu \frac{\sinh \beta\mu B_0}{\sqrt{\cosh^2 \beta\mu B_0 - 2e^{-\beta J}\sinh 2\beta J}} \quad (1.36)$$

N is the number of spins and μ is the permeability. $M \to N\mu$ for $B_0 \to \infty$; this is the saturation of M. We see from eqn. (1.36) that $M = 0$ if $B_0 = 0$ and $T \neq 0$. We can calculate the partition function for large Systems with $B_0 = 0$ (*4*):

$$Z_N(T) = 2^N \cosh \beta J \text{ with } T \neq 0. \quad (1.37)$$

It can be shown (*18*) that the heat capacity is

$$C_B = k_B \frac{(\beta J)^2}{\cosh^2(\beta J)}. \quad (1.38)$$

The magnetic susceptibility is

$$\chi(T) = \frac{\beta\mu^2\mu_0}{1 - \tanh(\beta J)}. \quad (1.39)$$

Furthermore we know that the internal energy for a $N \times N$ model is

$$E = -(N-1)J\tanh(\beta J). \quad (1.40)$$

The Hamiltonian of the *two–dimensional* Ising model is

$$\mathcal{H} = -\sum_{\langle i,j \rangle} s_i s_j - H \sum_i s_i. \quad (1.41)$$

This model has been solved on a square lattice It has been found that $T_C = 2.2692\, k_B^{-1}$ (*18*). The magnetization function is:

$$M_C(T) = \begin{cases} (1 - \sinh^{-4} 2\beta J) & \text{if } T < T_C \\ 0 & \text{if } T > T_C \end{cases} \quad (1.42)$$

We do not know the general solution for the two–dimensional Ising model with arbitrary external magnetic field. The magnetization is the order parameter in these systems. Below the known critical temperature we expect *ferromagnetism*; above T_C *paramagnetism* is observed. The analytic solution for the three–dimensional Ising model is still subject

Statistical mechanics – a short review

to research. The ground state of both models described so far is two times degenerate. It consists of the spin configuration with all spins being aligned parallely in one direction. This direction is dependent upon the external magnetic field.

If $J < 0$ we are dealing with the *antiferromagnetic* case. Antiparallel alignment of the spins is thus preferred. Hence we find that the ground state is a checkerboard (1). In the case of a vanishing external magnetic field we will get the same energy and therefore heat capacity as in the ferromagnetic case. Because of the antiferromagnetic coupling we get of course different M and c. If we have a *bipartite* lattice (i. e. it can be divided into two sublattices A and B: Site A has only B neighbors and vice versa) we can consider this two sublattices separately.[11] For next neighbor coupling we can define new spins (1)

$$s'_i = \begin{cases} +s_j & \text{if } j \in A \\ -s_j & \text{if } j \in B \end{cases} \quad (1.43)$$

Since $s'_i s'_j = -s_i s_j$ introducing new spins changes the spin of J and we retain the results of the ferrmagnetic case. If $H \neq 0$ then H has to be reversed on the B lattice. So we obtain the thermodynamic properties of the ferromagnet if we switch the sign of J and introduce the *staggered field* $H_A = H$ and $H_B = -H$. The problem may be tackled using variational methods for the sublattices. We find two competing states in the phase diagram. The ferromagnetic state with $m_A = m_B$ and the antiferromagnetic state with $m_A = -m_B$ (1).

---

[11]A square lattice for instance is bipartite, we are only considering this lattice type here.

Chapter 2.

Monte Carlo simulation

(*3*) In this chapter we'll explain the general principles of *Monte Carlo methods*. Of course we can only offer a concise presentation. Our motivation for this chapter is our desire to solve the two-dimensional *Ising model*. Doing this analytically is a tour de force (*19*) so we'll explore a different method. The applications of Monte Carlo methods, however are not restricted to the realm of statistical mechanics, rather they are used in *all* thinkable scientific disciplines. For an excellent survey see(*13*, *17*). Monte Carlo methods all have in common their reliance on sequences of random numbers to perform the simulation. E. Fermi and N. Metropolis are thought to have introduced term *Monte Carlo method* while working on nuclear weapon projects in the Los Alamos National Laboratory. The very idea of Monte Carlo methods goes back to a problem posed by GEORGES-LOUIS LECLERC DE BUFFON, the *Buffon's needle problem* (*17*).

> Suppose we have a floor made of parallel strips of wood, each the same width, and we drop a needle onto the floor. What is the probability that the needle will lie across a line between two strips?

The probability could be calculated by performing the experiment thus obtaining π.

2.1. Fundamentals

One reason to rely on Monte Carlo simulation is that they give better result for high-dimensional integrals than the standard integration schemes. Using traditional integration we want to evaluate the one–dimensional integral:

$$I = \int_0^1 f(x)\mathrm{d}x \tag{2.1}$$

We would partition the interval $[0,1]$ into N slices of width $d = (1-0)/N$. The l-th slice has the area $f[(x_l + x_{l+1})/2]$. Now we would calculate:

$$I \approx \sum_{l=0}^{N-1} d \cdot f[x_l + x_{l+1}/2] \tag{2.2}$$

For a sufficient number of slices we'll get reasonable results for the integral. It can be shown (21) that the sum in the above equation converges to I with an error $\propto N^{-\kappa/d}$. κ is a constant for example $\kappa = 4$ for *Simpson's rule* and d is the dimension of the integral. This error arises because all dimensions are treated independently; i.e. we are effectively computing d one–dimensional integrals. The statistical error for Monte Carlo methods is proportional to N^{-2}. For $d > 8$ Monte Carlo methods outperform standard integration as far as precision is concerned (12).

2.1.1. Crude Monte Carlo

The naive approach also referred to as *simple sampling* would be to choose a random number x_n with $n = 1, 2, \ldots N$ in the interval $[0,1]$ and then calculate I:

$$I = \frac{1}{N} \sum_{n=1}^{N} f(x_n) \tag{2.3}$$

For sufficiently large N we'll get satisfactory results. Note that the generalization to more dimensions is straightforward since we just need to chose d random numbers. It may be shown (12). that the statistical error of the Integral I is given by:

$$\delta I = \sqrt[2]{\frac{\operatorname{Var} f}{N-1}} \quad \text{with} \quad \operatorname{Var} f = \langle f^2 \rangle - \langle f \rangle^2 \tag{2.4}$$

As we see from eqn. (2.4) that there is no dependence on the dimension. But the simple sampling method applied here suffers from a serious drawback: Uniform sampling of the interval does not make sense for functions with large variance in the interval. Areas carrying only little weight for the integral are sampled with equal probability as areas of great weight. This gives the reason for the term simple sampling. Thus the average

Monte Carlo simulation

of an observable $\mathcal{O}$ is calculated according to:

$$\langle \mathcal{O} \rangle = \frac{\sum_{i=1}^{M} \mathcal{O}_{n_i} \exp(-\beta E_{n_i})}{\sum_{j=1}^{M} \exp(-\beta E_{n_j})} \quad (2.5)$$

This is in general not the optimal algorithm so we'll introduce a more refined one in the next section.

2.1.2. Markov Chain Monte Carlo

As a preparation for the physical applications of our algorithm we will introduce a modification of the simple sampling introduced above. In statistical physics we are often interested in simulating a (thermodynamic) system. Instead of generating a new state for each step of the calculation we prefer the easier way to slightly and randomly change the state of the system. For reasons that will become clear later methods this approach is referred to as *Markov chain Monte Carlo* or MCMC. First we need to smoothen the function that we want to integrate on a given interval. To this end we rewrite I ([20]):

$$I = \int_0^1 f(x)\mathrm{d}x = \int_0^1 W(x)g(x)\mathrm{d}x \quad \text{with} \quad g(x) \stackrel{def}{=} \frac{f(x)}{W(x)} \quad (2.6)$$

$W(x)$ is a distribution function that has to satisfy:

$$\int W(x)\mathrm{d}x = 1 \quad (2.7)$$

Which means that $W(x)$ is a probability function. $W(x)$ has to mimic changes in $f(x)$ so that I converges much faster.

$$I \sim \frac{1}{N} \sum_{i=1}^{N} \frac{f(x_i)}{W(x_i)} \quad (2.8)$$

As explained above we generate new states from old ones. We use the *Metropolis* algorithm ([8], [16]) for this. This algorithm will be introduced in some detail later applied to the Ising model ([2]). Using the canonical ensemble we can write:

$$\langle \mathcal{O} \rangle = \int \mathcal{O}W(x)\mathrm{d}x \quad (2.9)$$

We'll use this equation to evaluate the integral. The basic idea is to generate random values in the specified domain which is here $[0, 1]$ and then construct new values from the old ones by adding (or subtracting) suitable random values. The new values are rejected with a certain probability. If rejected the old value is taken. The following procedure has to be conducted (7, 20):

1. Randomly select an initial value x_0 in the specified domain
2. Generate a random number η such that $0 < \eta < 1$
3. $\Delta x = h(2\eta - 1)$
4. $x \leftarrow x_0 + \Delta x$
5. $p \leftarrow \frac{W(x)}{W(x_0)}$
6. Generate a random number r such that $0 < r < 1$
7. If $p > r$: $x \leftarrow x_0$. Else x does not change.
8. Go to 2.

Before implementing the algorithm we need to choose a step size h. Δx is sampled from a uniform distribution $[-h, h]$. If we choose a large h this entails Δx being sampled in a rather large interval reducing the acceptance rate. A small h on the other hand though guaranteeing reasonable acceptance rates leads to small steps and thus slow convergence. It is important to find a compromise between acceptance rate and step size. The parameter of the acceptance rate is $W(x)$. Z is a normalization constant following from eqn. (2.7). We can now calculate $\langle \mathcal{O} \rangle$.

$$\langle \mathcal{O} \rangle = \frac{1}{N} \sum_{i=0}^{N-1} O(x_{n_1 + n_0 i}) \qquad (2.10)$$

We cannot accept every data point of the Monte Carlo simulation due to the correlation of data. Therefore we wait n_0 steps before accepting a new value. The sampling method we used here is also referred to as *importance sampling* since we weigh values in the interval according to their importance, that is their contribution to the integral. To obtain the average of an observable $\mathcal{O}$ we just have to calculate the sum over the important states.

$$\langle \mathcal{O} \rangle = \frac{\sum_{i=0}^{M} \mathcal{O}_{n_i} p_{n_i}^{-1} \exp(-\beta E_{n_i})}{\sum_{j=1}^{M} p_{n_j}^{-1} \exp(-\beta E_{n_j})} \qquad (2.11)$$

Note that the total number of states may be greater than M since we are interested only in the contributing states. If we set $p_{n_i} = p_{n_j} = p = 1$ in eqn. (2.11) we get eqn. (2.5).

2.1.3. Markov chains

All *Markov chains*[1] share a common structure: Given a propability distribution π[2] on some configuration space S we want to generate random samples from π. We classify Monte Carlo methods as either *static* or *dynamic* (23).

1. **Static:** Monte Carlo methods generate statistically independent samples from π

2. **Dynamic:** Monte Carlo methods use a stochastic process[3] with state space S having π as its *unique* equilibrium distribution.

Dynamic Monte Carlo methods are usually simulated on the computer. After our system has reached equilibrium we may measure time averages. For a sufficient long run–time of the simulation we expect reasonable results as time averages should converge to π–averages. In practice we only interested in Markov chains with discrete state space (i. e. countable). This requirement is fulfilled in statistical mechanics. Two ingredients specify a Markov chain:

- The **initial distribution** α: α is a probability distribution on S with $\Pr$[4]$(X_0 = x) = \alpha_x$. α is given as a row vector.

- The **transition matrix:** $P \stackrel{def}{=} \{p_{xy}\}_{x,y \in S} = \{p(x \rightarrow y)\}_{x,y \in S}$

The transition matrix describes the probabilities for transition from one state to another; it has to fulfill the following criteria:

1. $\sum_j p_{ij} = 1$

2. $0 \leq p_{ij} < 1$

---
[1] named after the Russian Mathematican ANDREY ANDREYEV MARKOV (1856-1922).
[2] Or more general in fancy mathematical language: a propability measure
[3] This process needs to be found which is often a non–trivial task
[4] This denotes the probability of being in state x at time $t = 0$.

One important property of Markov chain is the *Markov property*: *Successive transitions are independent*. It follows than that a Markov chain is completely determined by:

$$\Pr(X_0 = x_0, X_1 = x_1, X_2 = x_2, \ldots, X_n = x_n) \stackrel{def}{=} \alpha_{x_0} p_{x_0 x_1} p_{x_1 x_2} \cdots p_{x_{n-1} x_n} \quad (2.12)$$

n steps after we've been in state x we are in state y with probability $p_{xy}^{(n)}$.

$$p_{xy}^{(n)} = \Pr(X_{t+n} = y | X_t = x) \quad (2.13)$$

2.2. Monte Carlo simulations in statistical mechanics

2.2.1. Ergodicity and detailed balance

Now we return to our initial task—we want to simulate a stochastic process in statistical mechanics. If we want to perform a realistic simulation, every state of our system should be accessible from any other state after a sufficiently long time. This is necessary for generating a correct Boltzmann distribution. More formally:

$$\forall (x, y) \in S \, \exists n \geq 0 : p_{xy}^{(n)} > 0. \quad (2.14)$$

This mathematical condition is referred to as *irreducibility* in mathematical literature. In the physics literature the term *ergodic* Markov chain is far more common, we'll follow this tradition.[5] Since we are dealing with equilibrium statistical mechanics we are interested in a system in equilibrium. Thus we have to guarantee that the stochastic process will eventually reach the *unique* equilibrium distribution. To this end we impose the condition of *detailed balance*: The probability of being in state x and moving to state y is equal to the probability of being in state y and moving to state x. Or more formally:

$$\pi_i p_{ij} = \pi_j p_{ji}. \quad (2.15)$$

---

[5] In the mathematics literature ergodic is used as a synonym for irreducible referring to Markov chains with *finite* state space. Regarding to Markov chains with infinite state space irreducible and ergodic are *not* used synonymously (23).

Monte Carlo simulation

This is a weaker conditions than $p_{ij} = p_{ji}$ it is however stronger than the condition of *stationarity*. The latter requires only:

$$\forall y \in S : \sum_x \pi_x p_{xy} = \pi_y. \qquad (2.16)$$

This does not guarantee the uniqueness of the equilibrium distribution ([17]).

2.2.2. Acceptance ratios

As already mentioned finding the right Markov process generating samples according to a given probability distribution can be hard work. But if even we do not posses the knowledge of the transition matrix we can still sample states with a certain probability distribution. To this end we introduce *acceptance ratios*. We can rewrite the probabilities as

$$p_{xy} = g_{xy} A_{xy}. \qquad (2.17)$$

Here g_{xy} denotes the probability of a selection of a new state y from state x. The transition is accepted with a probability A_{xy}, the acceptance rate or acceptance probability. We can now write:

$$\frac{p_{ij}}{p_{ji}} = \frac{g_{ij} A_{ij}}{g_{ji} A_{ji}} \qquad (2.18)$$

The ratio A_{ij}/A_{ji} can take any value between zero and infinity which is why we can choose any g_{ij} and g_{ji} and then choose an apt acceptance ratio. An ideal Monte Carlo algorithm has acceptance ratios of one since all states are sampled according to the desired distribution, i.e. $A_{ij} = A_{ji}$. In this case, of course there is no reason to introduce acceptance ratios. So for a *real* algorithm the ratio A_{ij}/A_{ji} should be close to unity. In the next subsection an algorithm with acceptance ratios close to unity is presented.[6]

2.2.3. Metropolis algorithm

In the last section we developed the formalism of Markov chains as well as the conditions required for Monte Carlo processes in statistical mechanics. Metropolis *et al.* developed

---

[6] As we will see later there are configurations of the system that are sampled very inefficiently and other algorithms are to be preferred then.

a method used to generate transition matrices satisfying the detailed balance condition (16). This approach was generalized by Hastings later (8). We'll focus on a special case commonly referred to as *the Metropolis* algorithm; it is only a special case though an interesting one. The condition of detailed balance may be written as:

$$\frac{\pi_i}{\pi_j} = \frac{p_{ij}}{p_{ji}} \qquad (2.19)$$

In the canonical ensemble in statistical mechanics we have

$$\pi_i = \frac{\exp(-\beta E_i)}{\sum_j \exp(-\beta E_j)}.$$

And thus: $\pi_i/\pi_j = \exp(-\beta \Delta E)$ with $\Delta E \stackrel{def}{=} E_i - E_j$. We still have the freedom tho choose an appropriate p_{ij}. It can be shown (23) that the transition probability has to satisfy $p(x)/p(1/x) = x \forall x$ with $\stackrel{def}{=} \exp(-\beta \Delta E)$. The particular choice of Metropolis et. al was $p(x) = \min(1, x)$. We obtain the following conditions for acceptance or rejection:

$$p_{ij}(x) = \begin{cases} 1 & \text{if } \Delta E \leq 0 \\ \exp(-\beta \Delta E) & \text{if } \Delta E > 0 \end{cases} \qquad (2.20)$$

We already explained the Metropolis algorithm applied to the problem of numerical integration. The steps required for the Metropolis algorithm applied to the Ising model are:

1. Choose a an initial state

2. Choose a random spin at position i on the lattice

3. Calculate the energy change ΔE caused by a spin–flip of the spin at site i

4. Flip spin with probability $p_{ij}(x)$, see eqn. (2.20)

5. Go to 2

2.2.4. Initialization bias

In the beginning of our simulation the system is out of equilibrium. As long as the system is out of equilibrium our data will have a systematic error since the equilibrium distribution was not yet reached. After a time τ_{eq} the system is in equilibrium. Unfor-

Monte Carlo simulation

tunately in general we do not know this time. One possibility is to estimate the time is by plotting observables like magnetization, internal energy... as functions of time in one diagram. After a sufficient time we expect that the observables do not change. Now they have reached approximately equilibrium. But it may be just a local minimum. Therefore we'll perform at least two simulations with different initial conditions, e. g. $T = 0$ and $T = \infty$. We'll plot the respective observables of both simulations in one diagram. The cross-over point indicates the equilibration time. This has to be done for every observable, i. e. two observables in one plot since different equilibration times may exist[7] for different observables. There is still the danger that the system reached only metastability. Therefore it is advisable to perform simulations in the in these regions with different initial values and test the consistency. Most Monte Carlo simulations simulations display metastability near a first-order transition (23). If we know the equilibration time we'll discard date performed before equilibration was achieved. The statistical errors associated with an out-of-equilibrium system are of order τ/n as opposed to statistical errors after the initial transient being of order $(\tau/n)^{1/2}$.

2.2.5. Autocorrelation in equilibrium

Since Monte Carlo methods sample new states from old states there is a high correlation between subsequent states. To get rid of this correlation we have to wait a time τ_{auto} before taking the next data point so the points are uncorrelated. This means that we have only $n/2\tau_{auto}$ independent data points if we conduct a run of length n.

To quantify the correlation between a given observable $\mathcal{O}$ at time t and the same observable at time $t + t_0$ the *autocorellation function* is used:

$$C_{\mathcal{O}}(t) = \frac{\langle \mathcal{O}(t_0)\mathcal{O}(t+t_0)\rangle - \langle \mathcal{O}(t_0)\rangle\langle \mathcal{O}(t_0+t)\rangle}{\langle \mathcal{O}^2(t_0)\rangle - \langle \mathcal{O}(t_0)\rangle^2} \qquad (2.21)$$

Sometimes eqn. (2.21) is used without the normalization factor, this is the *autocovariance*. It follows from eqn. (2.21) that the correlation is positive if $\mathcal{O}(t)$ and $\mathcal{O}(t + t_0)$ fluctuate in the same direction. It is negative if they fluctuate in the opposite direction, they are anticorrelated. It may be shown that $C_{\mathcal{O}} \sim \exp(-t/\tau_{auto})$ (12). One possibility to obtain τ_{auto} once we know the autocorrelation function is to plot it as function of time on semi-logarithmic axes. The slope which can be obtained using a linear fit is τ_{auto}.

---

[7]In fact they do exist. τ_{eq} for E is typically smaller than for m (17).

2.3. Analyzing the data – T_c and more

Once the Monte Carlo simulation is finished the most important task has yet to be done. The data need to be analyzed. To achieve this aim we first have to find the critical temperature T_c. With this temperature we can calculate the critical exponents.

2.3.1. Finite size scaling and the Binder ratio

The basic idea is that in the vicinity of the critical temperature the correlation length ζ diverges; i.e. the typical size of clusters of correlated spins explodes. Due to the finite size of the lattice there can be no strictly mathematical divergence. Once the dimension of the system L is reached the correlation length is cut off. It can be shown (*21*) that observables (here m) near T_c and for large L can be written:

$$\langle m_L \rangle \approx L^{\beta/\nu} \tilde{M}\left[L^{1/\nu}(T - T_c)\right] \qquad (2.22)$$

$\tilde{M}$ is an unknown *scaling function*. From eqn. (2.22) we see that in the limit of large L data of systems with different L should cross at $T = T_c$. For this, however β/ν has to be known. Since in general we do not know the critical exponents we cannot use eqn. (2.22) directly. Instead the dimensionless fourth order *Binder cumulant* U_4 (*5*) is used. It is defined as:

$$U_4 = 1 - \frac{\langle m^4 \rangle}{3\langle m^2 \rangle^2} \sim \tilde{G}\left[L^{1/\nu}(T - T_c)\right] \qquad (2.23)$$

For $L \to \infty$

$$U_4 \to \begin{cases} \frac{2}{3} & \text{if } T < T_c \\ U^* \approx 0.61 & \text{if } T = T_c \\ 0 & \text{if } T > T_c \end{cases} \qquad (2.24)$$

If we know T_c we can calculate the critical exponents if we can plot the logarithm of the thermodynamic observables as a function of the logarithm of the absolute reduced temperature $\varepsilon = |1 - T/T_c|$ on logarithmic scales. Fitting the function to a straight line gives us the critical exponent as the slope of the linear fit function. For this method to be precise at least twenty points are required—hey all have to be in the critical region (*17*).

Monte Carlo simulation

Besides one problem is that we to have very precise data points which typically requires very long computations to get reasonable results using the Metropolis algorithm. As an alternative we can plot the logarithm of the observables against the logarithm of the system size L. From the linear fit we use the slope as the critical exponent.

Once we have determined T_c we can check whether we have the right value by plotting the fourth order Binder cumulant against $L^{\frac{1}{\nu}}(T - T_c)$. If we know T_c we can plot this cumulant for different values of ν. For the Ising model in two dimension the value is known: $\nu = 1$.

Chapter 3.

Selected results

If not stated otherwise temperature is in units of energy kT and thermodynamic observables are expressed per spin and per $|J|$. In the plots containing the magnetic field we set $\mu = 1$.

3.1. One–dimensional Ising model

The simulation[1] was performed without relaxation steps since they were not necessary. The number of MC steps per lattice site was 1000. All simulations were performed on a square lattice with periodic boundary conditions (see fig (B.1)).

Magnetization: Without external magnetic field $\langle |M| \rangle$ is vanishing for large T since fluctuations of the spins lead to a vanishing net magnetization. For lower temperatures however spins tend to align parallel because of positive next–neighbor coupling and therefore the magnetization increases. It is only at $T = 0$ that all spins are aligned parallel and therefore the maximal magnetization per spin is realized. If we add an external magnetic field we obtain higher values for the magnetization than without external field. We see from the plot that the magnetization with B decreases slower if $T \to \infty$ than without external field. The next–neighbor interaction is the reason for this behavior (10, 15). We can also see from the plot that the data we obtained from our simulation are in close agreement with theoretical results (11).

Energy: The energy has been plotted for $B = 0$ and for $B \neq 0$. The results have been compared with the analytical results and we get the same result. We see that the

---
[1] Nicer results produced using *Mathematica* can be found in (15). The interpretation however remains the same.

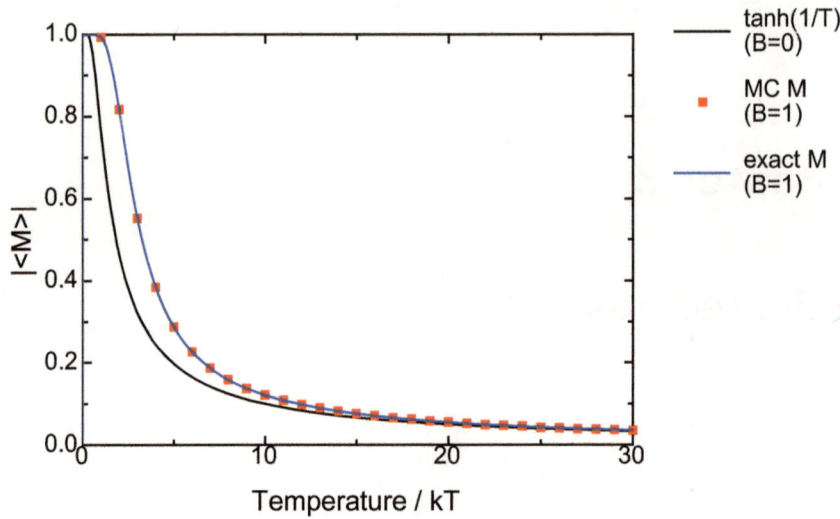

Figure 3.1.: Magnetization 1D Ising model theoretical and MC results

absolute value of the energy per spin goes to one as $T \to 0$. The phase transition occurs only at $T = 0$. This is not the case for the two–dimensional ferromagnetic Ising model. If we add a magnetic field we can observe that the energy graph is shifted to higher energies (if we increase B). For stronger magnetic fields it approaches a linear function with saturation at $= -1$ for higher temperatures.

Susceptibility: Comparing the case $B = 0$ to $B \neq 0$ we see that an external magnetic field reduces the peak of χ. For infinite systems the peak is a divergence. The magnetic field however is a constraint on the spins so that the peak is shrinking if B is increased. Since χ is formally a variance we can understand from the effect of increasing B on M that also χ has to become smaller if we increase B.

The same arguments can be used for the **heat capacity** if we substitute E for M and χ for c. Increasing B shifts the peak of c to higher temperatures and broadens it as we've written above.

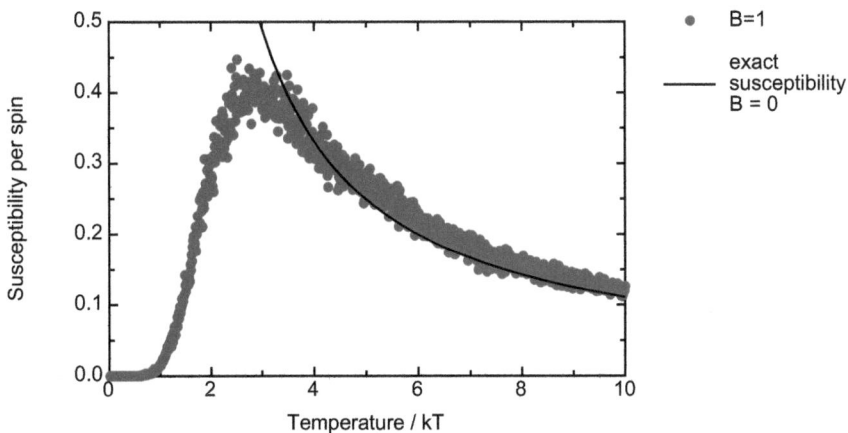

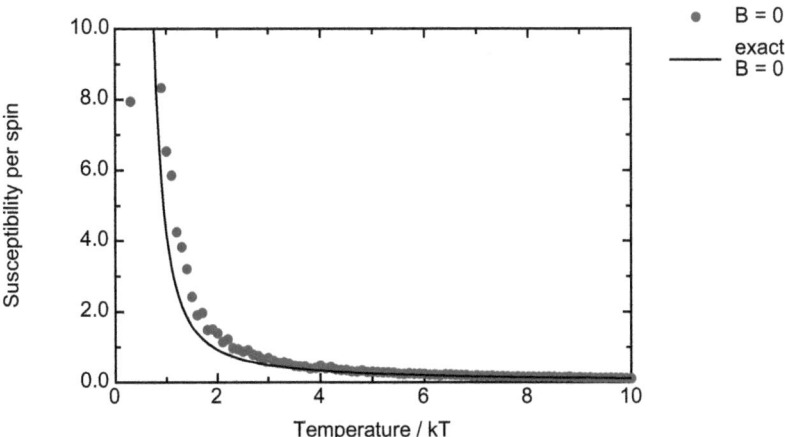

Figure 3.2.: Magnetic susceptibility for $B = 0$ and $B = 1$ theoretical and MC results

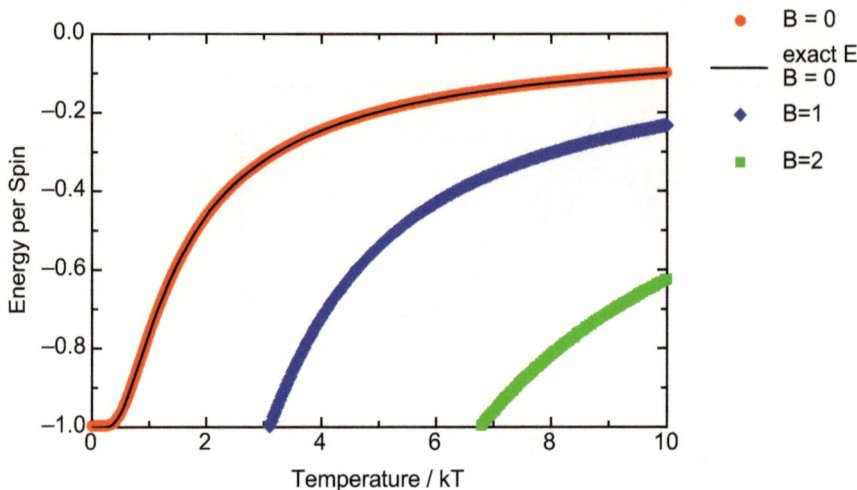

Figure 3.3.: Energy 1D Ising model theoretical and MC results

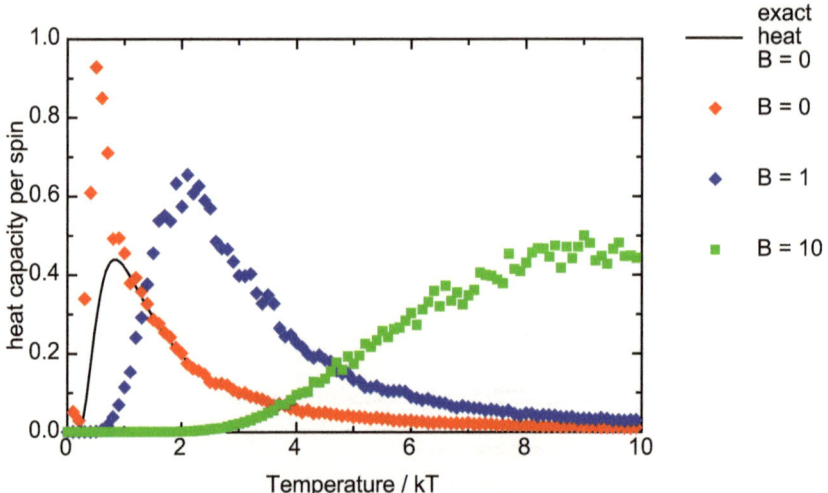

Figure 3.4.: Heat capacity for $B = 0, 1 2$ and $B = 10$ from MC simulation and exact result without field

3.2. Two–dimensional Ising models

We used a 100×100 lattice if not stated otherwise and 10000 MC steps per lattice site. For the simulation of the susceptibility we additionally had 100 relaxation steps in the critical region between the measurements.

3.2.1. The ferromagnetic Ising model

Equilibration time

We already mentioned in the second chapter the importance of the equilibration time. Now we can estimate it for E and M. So as to achieve this end E and M have been plotted for different start temperatures: $T = 0$, i.e. all spins parallel and $T = \infty$, i.e. a random spin configuration. This simulations were performed at $T = T_c$.

Different observables in one plot: If we plot E and M both with $T = \infty$ against the number of MC steps we see that it takes approximately 2000 MC steps for E to reach equilibrium. The magnetization has not really reached equilibrium even after 10000 MC steps. This is as we'll see one of the problems of the Metropolis algorithm. The equilibration time is of course smaller for regions outside the critical region.

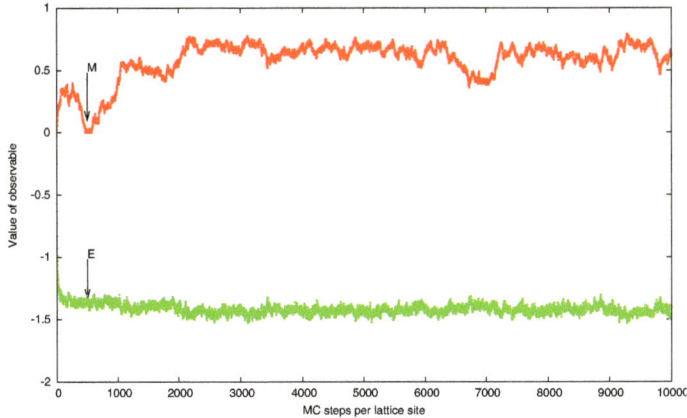

Figure 3.5.: E and M as a function of MC steps with initial configuration $T = \infty$

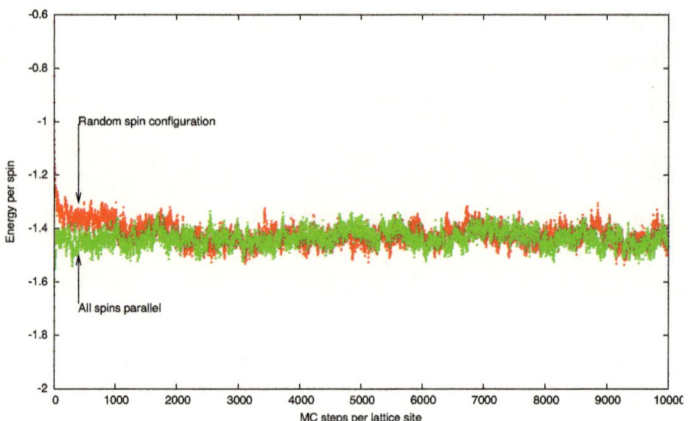

Figure 3.6.: E in two different initial spin configurations as a function of MC steps

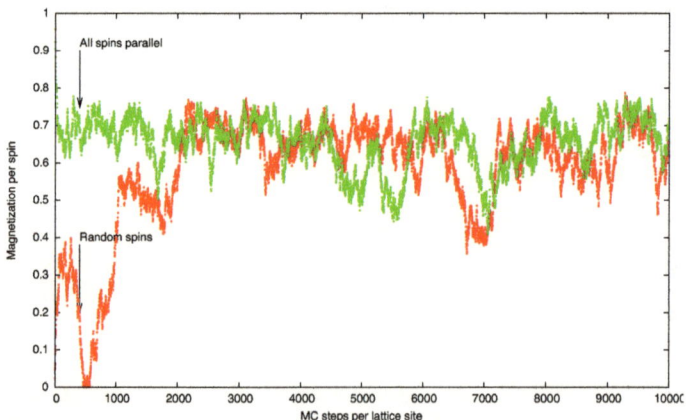

Figure 3.7.: M in two different initial spin configurations as a function of MC steps

Selected results

Magnetization: We plotted the magnetization for two different initial spin configuration against the number of MC steps. From the plot we see that it takes at least 10000 MC steps to reach equilibrium. Even after this 10000 steps M is still subject to substantial fluctuations.

Energy: We already saw that the fluctuations of the energy are much weaker than the fluctuations of M. From the energy–MC steps plot we see that not more than 3000 MC steps are required so that it is hard to distinguish between the two energy curves.

Finite size effects

Magnetization:[2] It displays a sharp increase below a certain temperature[3]. Above the critical temperature the magnetization is vanishing. If we go to lower temperatures, however the magnetization goes to 1 as we approach $T = 0$. At this temperatures all spins are aligned parallel in one direction. For smaller systems we can see the fluctuation especially in the critical region where the fluctuation are generally high and for large T. In this region random fluctuations increase fluctuations due to small lattice size. Furthermore with growing lattice site also the transition from vanishing magnetization to non-vanishing magnetization becomes sharper. This is a sign for a phase transition (*10*).

Energy: Its behavior is a further hint at a phase transition. As we approach the critical temperature (that we do not know exactly yet) there is a larger increase in E. Also different lattice size do not differ in E now. The differences between lattices with different L, however, became apparent if we consider the behavior above T_c. With decreasing L the energy curve approaches a linear function in the high temperature regime.

From m we can derive the **magnetic susceptibility:** The susceptibility has a sharp peak at approximately $kT = 2.28$; for lower or higher temperatures it falls very fast. In this region we assume to find a phase transition. The peak indicates fluctuations of m in this region. Smaller lattices do not exhibit a sharp peak since the cut–off of the diverging correlation length occurs early owing to the small size of the system.

---

[2] As above we refer here to $\langle |M| \rangle$ as magnetization.
[3] which is the critical temperature, as we'll see.

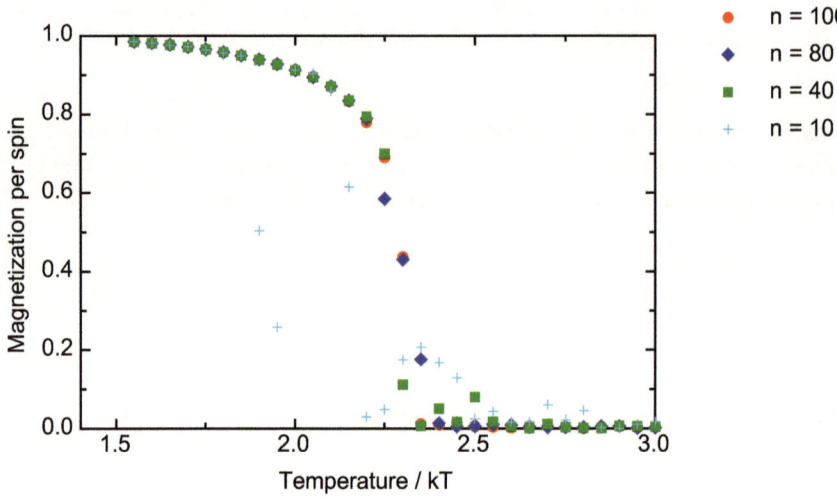

Figure 3.8.: Magnetization for different lattice sizes

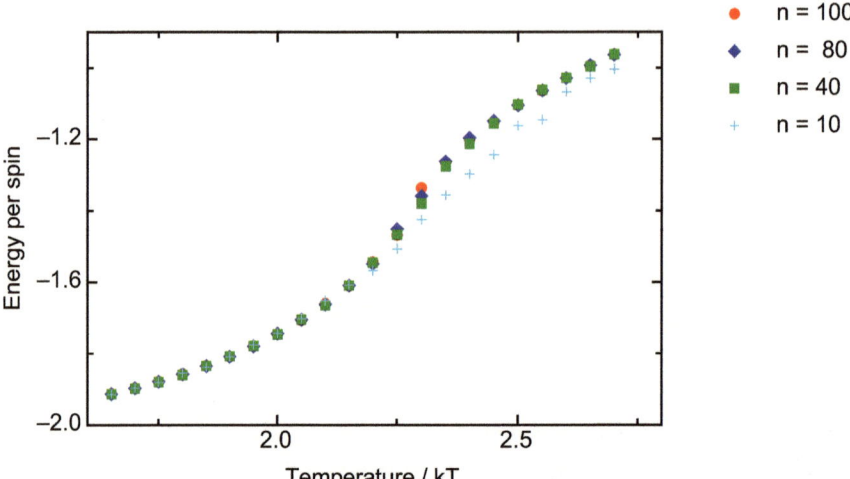

Figure 3.9.: Energy for different lattice sizes

Selected results

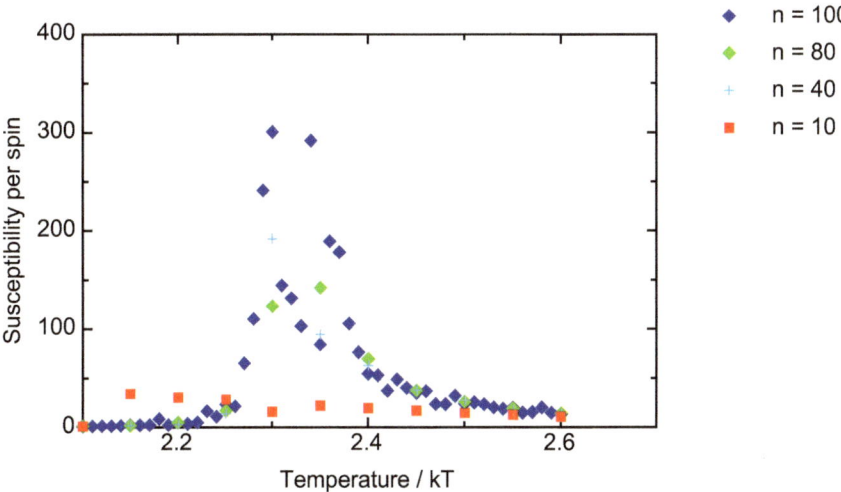

Figure 3.10.: Magnetic susceptibility for different lattice sizes

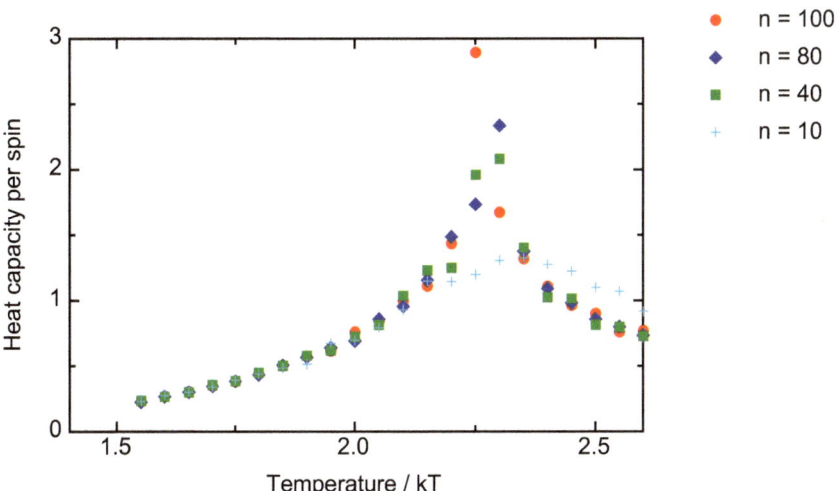

Figure 3.11.: Heat capacity for different lattice sizes

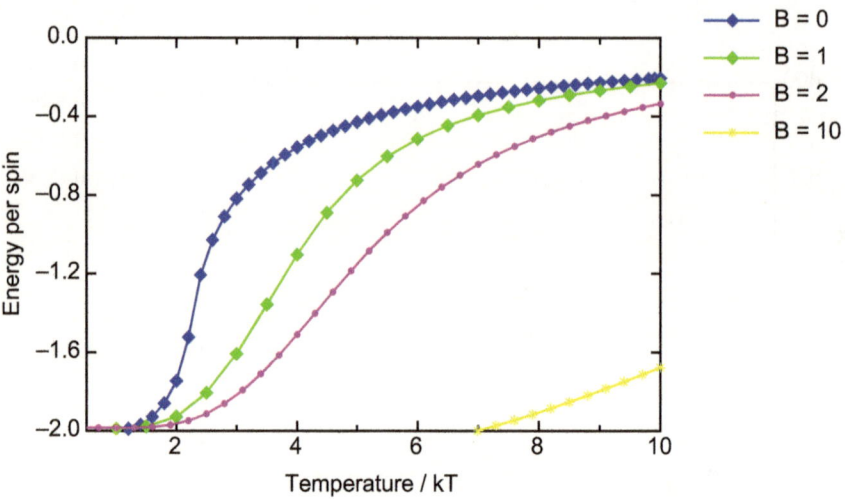

Figure 3.12.: Energy of 2D Ising ferromagnet for different B

Heat capacity: c has a shape similar to the magnetic susceptibility. The peak of c, however, is weaker than in the case of χ. The same finite size effects as for χ are observed.

Ferromagnetic Ising model with magnetic field

Energy: We get basically the same results as for the 1D case. For low temperatures the energy saturates at $E = -2$. However, if the external magnetic field is increased the energy curve approaches a linear function and is shifted to higher temperatures. This happens for the same reasons as explained for the 1D model: Because of the external field spins align parallel. This means that the phase transition to the ferromagnetic state occurs at higher temperatures in the presence of an external field.

Magnetization: The magnetization saturates at $M = 1$ below the critical temperature. This critical temperature itself is changed through an external field as explained above. The magnetization follows the general pattern that we observed for the energy: An increase of B leads to a distortion of the magnetization; for $B \to \infty$ a M approaches a parallel to the x-axis.

Selected results

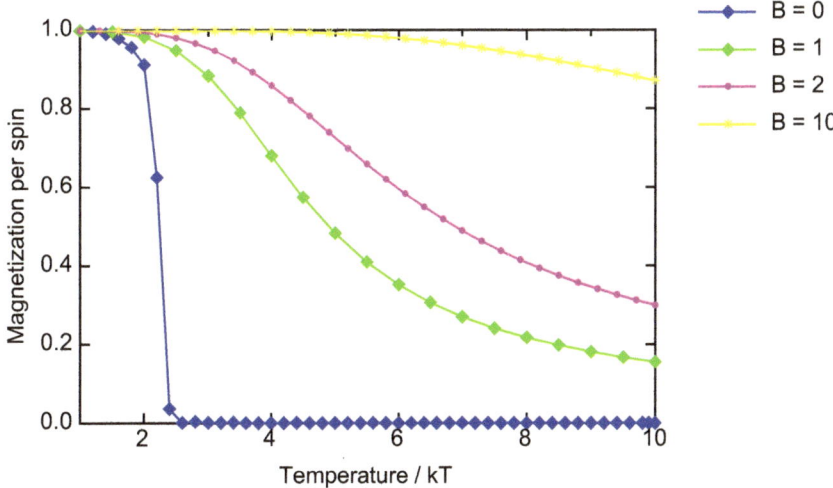

Figure 3.13.: Magnetization of 2D Ising ferromagnet for different B

Magnetic susceptibility: The magnetic susceptibility displays a very sharp peak approximately at the critical temperature. The presence of an external magnetic field shifts the susceptibility to higher temperatures. This of course happens because the phase transition occurs at higher temperatures. The peak is broadened since the external field causes the magnetization to become more linear (see magnetization). Since there is no sharp transition from the paramagnetic state to the ferromagnetic in M, χ is very small. The fact that χ is very large if $B = 0$ maybe partially explained by appealing to the low number of relaxation steps taken between subsequent measurement.

Heat capacity: The heat capacity has a peak at T_c. The peak is shifted right if $B > 0$ (see χ for explanation).

3.2.2. Antiferromagnetic Ising model

Antiferromagnetic Ising model with $B \neq 0$

Magnetization: The magnetization is vanishing for all temperatures if $B = 0$. Due to thermodynamic fluctuations for very large T, very small magnetizations may occur. This behavior may be explained appealing to the antiferromagnetic ordering. Owing

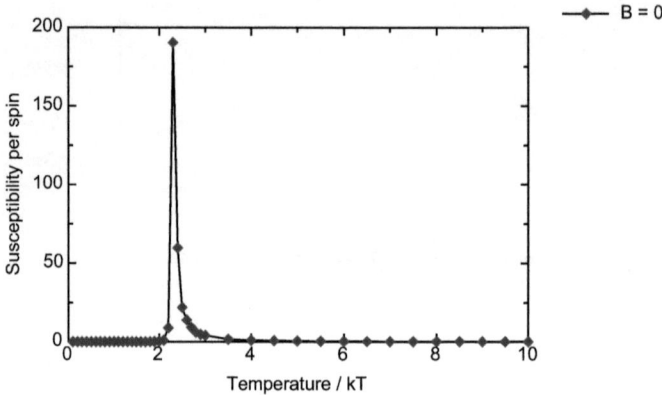

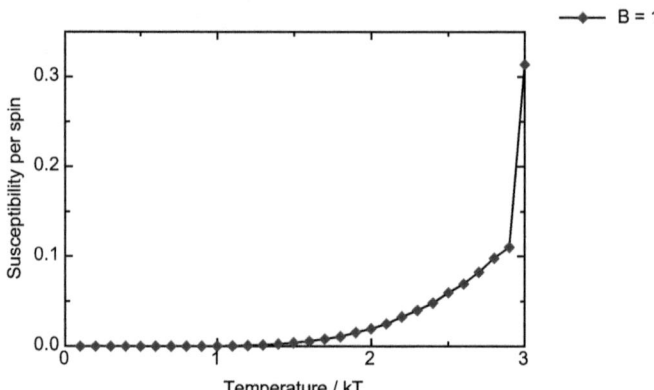

Figure 3.14.: Susceptibility of 2D Ising ferromagnet for different B

Selected results

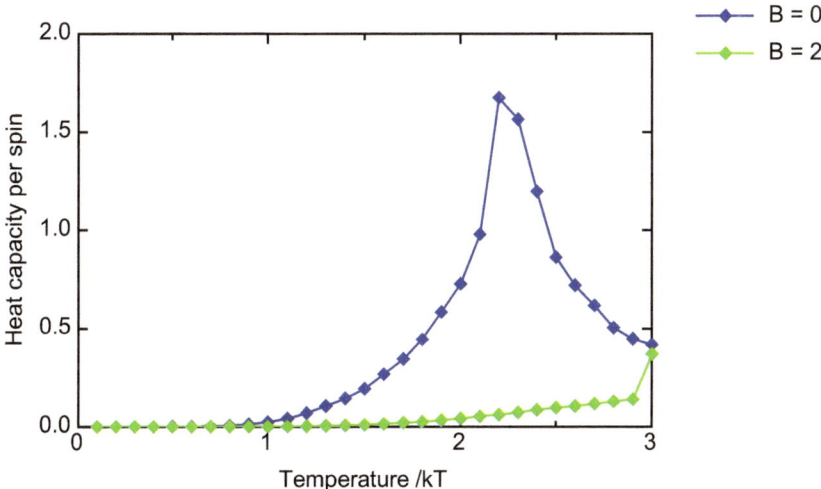

Figure 3.15.: Heat capacity of 2D Ising ferromagnet for different B

to this ordering spins tend to align antiparallel. Because of the very definition of M it follows that M should be very small. Since we are interested in $\langle |M| \rangle$ we have a vanishing magnetization that can be only zero if the number of up-spins is the same as the number of down-spins. This holds only for $B = 0$. In the presence of a magnetic field we have a critical magnetic field B_c that leads to antiferromagnetic ordering. And above this field the ferromagnetic ordering prevails. From the plot we see that the critical field is at $B = 4$. Increasing the external magnetic field slightly above B_c we see that $M \to 1$ if $T \to 0$. This result holds only for the square lattice. Below the critical field antiferromagnetic order is realized and we observe in the limit of low temperatures zero magnetization. Since we have a perfect antiferromagnetic order if $T = 0$ all spins are antiparallel. Therefore the magnetization vanishes.

Energy: The energy, however, of the ferromagnetic and of the antiferromagnetic Ising model in two dimensions cannot be distinguished. We already explained this based on a mathematical argument. This only holds if $B = 0$ otherwise we get different results. If the external magnetic field is increased the energy curve is shifted to higher total energies. This behavior holds up to the critical magnetic field of $= 4$ for the square lattice. For higher magnetic fields we get a different behavior. Above $B = 4$ the magnetic Energy approaches positive energies. For $B > 4$ the system is ferromagnetic.

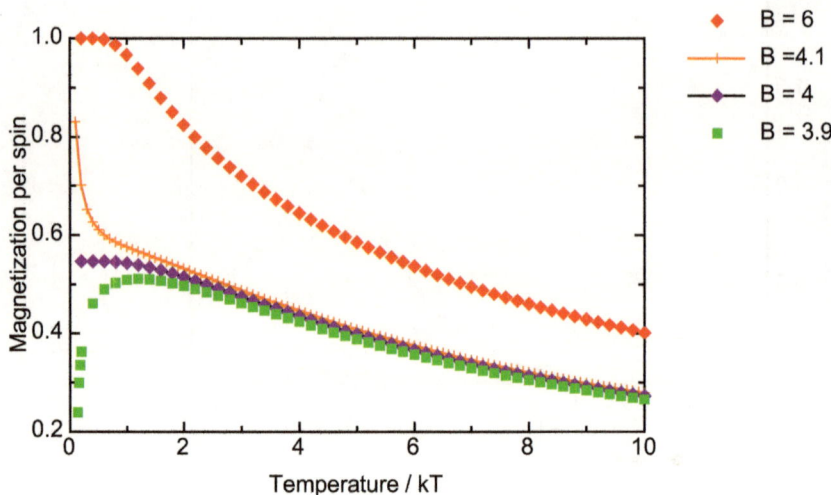

Figure 3.16.: Magnetization of 2D Ising antiferromagnet for different B

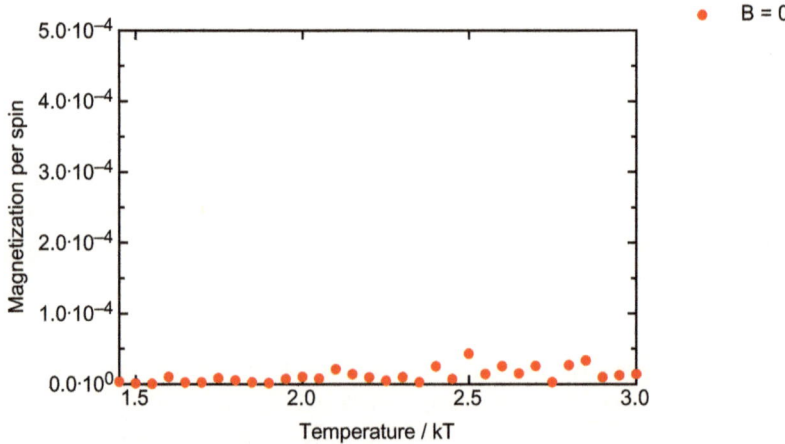

Figure 3.17.: Magnetization of 2D Ising antiferromagnet for $B = 0$

Selected results

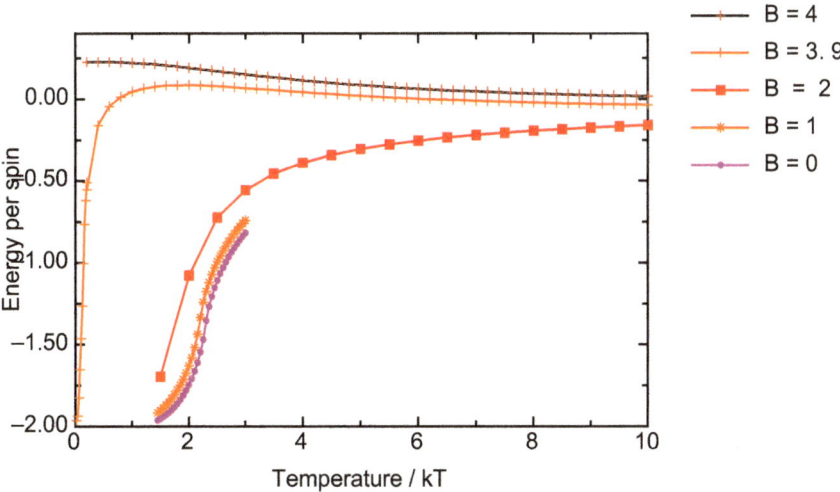

Figure 3.18.: Energy of 2D Ising antiferromagnet for different B

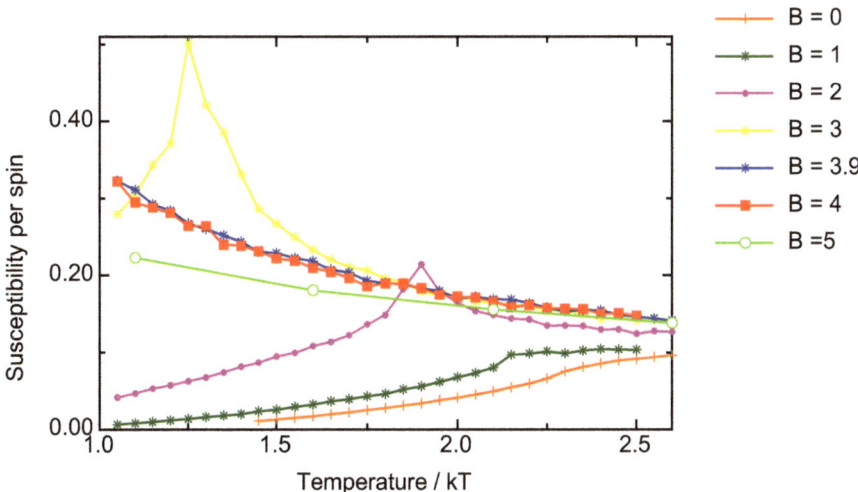

Figure 3.19.: Susceptibility of 2D Ising antiferromagnet for different B

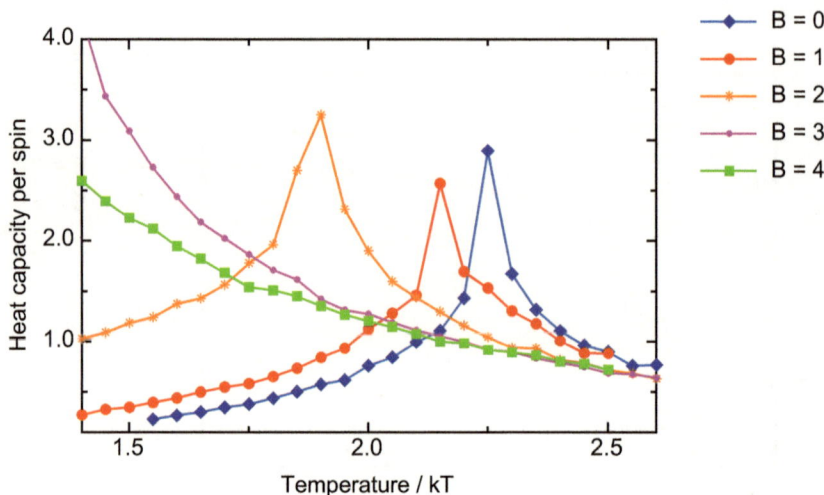

Figure 3.20.: Heat capacity of 2D Ising antiferromagnet for different B

Susceptibility: Since the susceptibility is calculated from the magnetization we expect similarities with the magnetization. If the magnetization is increased the peak of the susceptibility is shifted to the lower temperature region. As long as $B < 4$ the system is in the antiferromagnetic state. Above $B = 4$ however the system display ferromagnetic order. In the latter case a further increase of B reduces the susceptibility as we saw (see ferromagnetic Ising model).

Heat capacity: The heat capacity is calculated from the energy as mentioned earlier. Therefore we get the expected result that we know from the Ising ferromagnet. For $B \neq 0$ however we find different results. If we have a weak magnetic field we find that the peak of the heat capacity is shifted to lower temperatures if we increase B. Also the curve is broadened. For $B = 4$ the peak is already shifted to temperatures below $1\,\text{kT}$.

3.3. The phase transition

If we look at the magnetization and at the susceptibility of the ferromagnetic Ising model we can see a striking feature near $T = T_c$ that is around $2.2\,\text{kT}$. There is a continuous phase transition. We can understand the behavior of the observables by looking at the spin configuration at different temperatures. At sufficiently high temperatures spins

Selected results

are uncorrelated and random. As the temperature is lowered however the spins tend to align parallel due to the positive coupling constant that encourages this type of order. The characteristic feature of a phase transition is the divergence of the correlation length. In the ferromagnetic Ising model clusters of parallel aligned spins with all spins either pointing up or down begin to form. Their size is given by the correlation length. At $T = T_c$ the size can grow very large with the upper bound given by the system size. Below T_c the majority of spins point in either up or down direction. The chosen direction is dependent on the thermodynamical details of the configuration. Lowering the temperature increases the size of the clusters so that in the limit of $T = 0$ all spins point into one direction and therefore the magnetization per spin is ± 1.

One problem of the Metropolis algorithm has become apparent from the results of our simulation. In the vicinity of T_c critical fluctuation occur. This belongs to the *critical phenomena*. In this region large clusters contributing massively to E and M a flip of the orientation of one cluster results in a great change of E and M. The fluctuations are even stronger in the derived observables χ and c that are obtained from E and M. In this region far more MC steps are required to get an acceptable result as we need it to calculate T_c and the critical exponents for instance. To tackle this problem other algorithms have been developed. Normally *cluster algorithms* are used. They flip whole clusters consisting of spins of equal orientation. By flipping clusters instead of single spin the problem of the fluctuation is solved (24).

3.4. The Critical temperature and critical exponents

3.4.1. Binder cumulant

The fourth order Binder cumulant (6) was calculated for different lattice sizes with 10^5 MC steps per lattice site without relaxation steps. From the plot we can see that there is more than just one cross–over point. The exact critical temperature is between two such points at $T_c = 2.2692\,\text{kT}$. Since we did not perform any relaxation steps we get a result of approximately 2.27 for T_c.

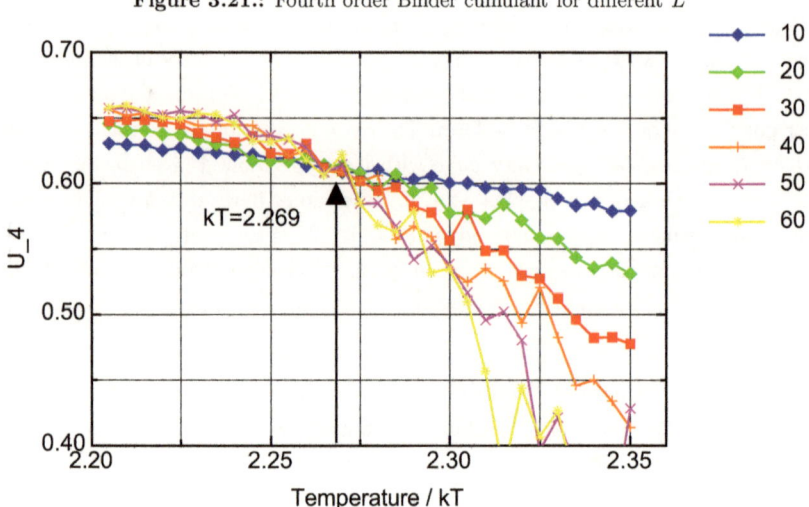

Figure 3.21.: Fourth order Binder cumulant for different L

3.4.2. Critical exponents

We plotted the $\log \mathcal{O}$, for $\mathcal{O}$ an observable, against $\log L$ at $T = T_c$ and we performed a linear fit. The slope of the linear function is the exponent. For the magnetization very big fluctuations occur so that a smaller system size had to be taken since the very reason for the fluctuations is the divergence of the correlation length. For smaller system the cut-off happens earlier—the slope however is different if we are to far from the critical region. Unfortunately since we did not perform enough relaxation steps in the critical region our result was to imprecise and inaccurate results for the critical exponents were the consequence. Normally a different algorithm is use to calculate the critical exponent or a more efficient language then Java. However even on a fast computer approximately 10^6 relaxation steps are required in the critical region additionally to 10000 MC steps that are required so as to reach equilibrium.

Once we have obtained T_c we can show that we have approximately at least the right value by plotting U_4 against $L^{\frac{1}{\nu}}(T - T_c)$. We see from the plot that the values of U_4 for different lattice sizes are approximately on one curve. Thus it follows that our value of T_c is right with a certain inaccuracy. For this we used $\nu = 1$.

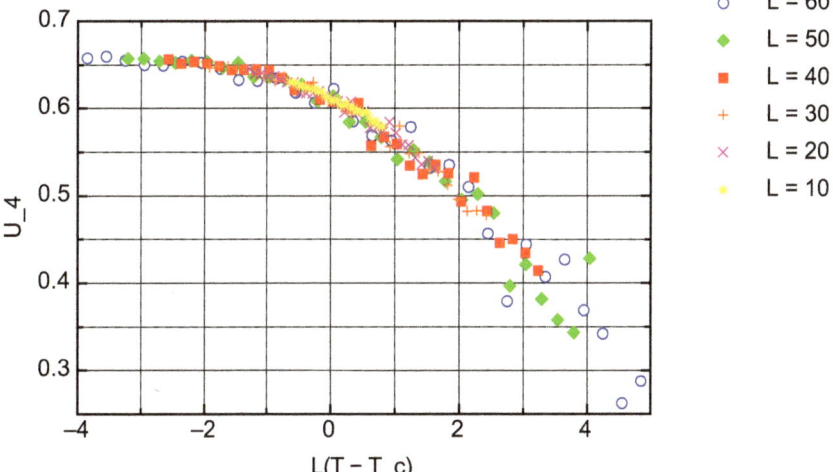

Figure 3.22.: Fourth order Binder cumulant against size–scaled reduced temperature

Appendix A.

Source code of MC integrators

A.1. Crude Monte Carlo

Listing A.1: Source code for crude MC integrator

```java
/**
 * @author Michael Adler
 * @date 5/17/2010
 * This is an crude Monte Carlo integrator
 * using the IBAA rng. Integration is
 * performed in [0,1]*interval, so by adjusting
 * the double interval the domain of integration
 * can be changed. The function f is integrated.
 */
package de.Michael3600876;

import java.util.*;
/*Random number generator implemented by my brother*/
import de.alex4017769.IBAA;

public class CrudeIntegrator {

    public static void main(String[] args) {
        /* change domain if necessary */
        double interval = 1;
```

```
                Random r = new IBAA(( int ) System.
                   currentTimeMillis());
                /* # of iterations */
                final int ITERLEN = 10000;

                double rv = 0;
                for (int i = 0; i < ITERLEN; ++i) {
                        double x = r.nextDouble() * interval;
                        /*enter the function to be integrated
                           */
                        double f = Math.sin(1 / x);
                        rv += f;
                }
                System.out.println(rv / ITERLEN);
        }
}
```

A.2. Markov chain MC integrator

Listing A.2: Source code for Markov chain MC integrator

```
/**
 *
 * @author Michael
 * @date 05/17/2010
 *
 * This is a naive integrator using Markov
 * chain Monte Carlo, viz the Metropolis
 * algorithm and the IBAA rng in the
 * implementation of my brother Alex. The
 * function f is integrated over the domain
 * [0,1]*interval; see crude MC integrator.
 * The algorithm is described in "An Introduction
 * to Computational Physics, 2nd Edition" by
 * Tao Pang.
 */
package de.Michael3600876;
```

Source code of MC integrators

```java
import java.util.*;

import de.alex4017769.IBAA;

public class Integrator {
        static Random r = new IBAA((int) System.
            currentTimeMillis());

        /* middle of [0,1)*/
        static double stepsize = 0.4;

        /* length of interval */
        static double interval = 1;
        /* initial value x_0 */
        static double x = r.nextDouble() * interval;

        static double f;

        /* # of accepted displacements */
        static int acceptno;

        static int skiprate = 30;

        /* # of iterations */
        static final int ITERLEN = 100000;


        public static void main(String[] args) {
                for (int i = 0; i < ITERLEN; ++i) {
                        x = r.nextDouble() * interval;
                        /* Metropolis function */
                        metropolis();

                }
                System.out.println(f / acceptno);
```

```java
        // System.out.println("f = " + rv / ITERLEN);
}

private static void metropolis() {

        double oldx = x;

        double neww, rv;

        /* generates new x (new displacement) */
        generateNewX();
        if (p(x) > r.nextDouble()) {

                /* accept new state and calculate
                new weight: */
                neww = w(x);
                ++acceptno;

                /*enter here function */
                f += Math.sin(1 / x);
        } else {

                /* do not accept new step */
                x = oldx;
        }

}
/**
 *
 */
private static void generateNewX() {
        int i = 0;
        do {
                ++i;
```

Source code of MC integrators

```
                if (i % skiprate == 0) {
                        /* do not accept every value
                        because of correlation of
                        subsequent */
                        x += 2 * stepsize * (r.
                        nextDouble() - 0.5);
                }
        } while (x < 0 || x > interval);

}

/**
 * @return W(x) the weight of the function;
 cf. description in thesis.
 * */
private static double w(double x) {
        /* enter also here f */
        return Math.exp(Math.sin(1 / x)) - 1;
}

/**
 * @return G(x) see: description
 *in main text of thesis.
 * */

private static double p(double oldx) {
        return w(oldx) / w(x);
}

}
```

Appendix B.

Implementation of the Ising models

B.1. Skeleton of the code

The code of the Ising models is written in *Java* an object oriented language which however is not as efficient as *fortran* or *C*. It was used because the author learned this language. Only the two-dimensional model is described here because once it is understood also the easier one-dimensional model will be understood.

For the simulation of the Ising model *periodic boundary conditions* are used, they are illustrated in fig (B.1).

Certain parameters are user input they are highlighted in the source code if they are set the calculation may begin. Typically one has to set the initial temperature, the end temperature, the step size, the external magnetic field, the coupling constant, the lattice length[1], the number of Monte Carlo steps and the number of relaxation steps. Furthermore one can choose whether the initial spin configuration corresponds to $T=0$ or to $T=\infty$.

First the spin array `spins` is initialized (this is the `doInit` function in the code), i. e. it is filled either randomly or with all spins in one direction. From this the magnetization is calculated: $M = \sum_i s_i$ over the whole lattice. Then the energy is calculated according to:

$$E = -\frac{1}{2} J \sum_k \left(s_k \sum_{i \text{ n. n. to } k} s_i \right) - MB$$

---
[1]That is the lattice has length × length sites.

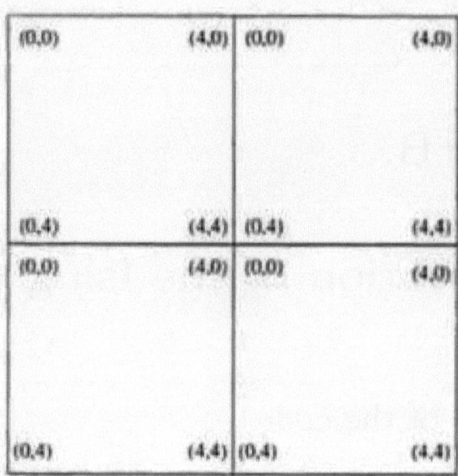

Figure B.1.: Illustration of periodic boundary conditions (*17*)

The actual calculation is performed in the `Metropolis` function that performs one MC step per lattice site. The function is called once for every lattice site times the number of `Cycles`. This is the number of steps required so that the system is in equilibrium and we have small fluctuations due to statistics. Then a position (`posx`, `posy`) is randomly chosen at the lattice. The energy difference between the old state without spin flip and the new with spin flip is then calculated (`calculateDeltaE`). Now if the Metropolis criterion is fulfilled the spin is flipped. So as to improve performance $e^{-\beta \Delta E}$ is once calculated for a given T and saved in an array `weight`. If the spin is flipped we have to update M and E. We only need to add the difference between the old an the new E and M to get the current value. If we take into consideration that only one spin is flipped and the other spins do not change the change of energy between the two states $E_\nu - E_\mu = \Delta E$ can be written as (*17*)

$$\Delta E = -J \sum_{\langle ij \rangle} s_i^\nu s_j^\nu + J \sum_{\langle ij \rangle} s_i^\mu s_j^\mu = -J \sum_{i \text{ n.n. to } k} s_i^\mu (s_k^\nu - s_k^\mu). \tag{B.1}$$

The sum on the right-hand side is only over nearest neighbors of the flipped spin k. The only spin to change is the flipped spin hence $s_i^\mu = s_i^\nu$. Only two possibilities exist: $s_k^\nu = +1 \Rightarrow$ after spin flip $s_k^\mu = -1$ It follows that $s_k^\nu - s_k^\mu = -2$. If $s_k^\nu = -1 \Rightarrow$

Implementation of the Ising models

after spin flip $s_k^\nu = +1$ and thus $s_k^\nu - s_k^\mu = +2$. From this we see that

$$s_k^\nu - s_k^\mu = -2s_k^\mu \tag{B.2}$$

If we plug this in eqn. (B.1) we can rewrite ΔE:

$$\Delta E = 2J \sum_{i \text{ n.n. to } k} s_i^\mu s_k^\mu = 2Js_k^\mu \sum_{i \text{ n.n. to } k} s_i^\mu \tag{B.3}$$

This exactly is calculated by `updateE`. `updateM` calculates ΔM from eqn. (B.2).

Once the specified number of iterations has been performed the variables (called accumulators) are calculated; i.e. $\langle E^2 \rangle$, $\langle E \rangle$, $\langle M \rangle$ and $\langle M^2 \rangle$. These (or the smart observables that can be obtained from E and M) are written in a CSV file. Near the critical temperature T_c subsequent states of the Ising model are highly correlated. To decorrelate them we need to wait a certain time between to measurements. This means we flip spins but we do not update M and E. If the boolean `doRelax` is `true` then `relaxtime` relaxation steps are performed between subsequent measurements.

B.2. Source codes of the Ising models

B.2.1. One dimensional Ising model

Listing B.1: Source code for 1D Ising model

```
/**
 * This is an implementation of the one-dimensional
 * Ising model without magnetic field using periodic
 * boundary conditions. Boltzmann's constant is set
 * one. Coupling constant j = 1. => Temperature has thus
 * dimension energy. Metropolis sampling wih one site
 * spin flip is used. Part of the optimizations is described in
 * M. Hjorth-Jensen, Lecture Notes on Computational Physics,
 * University of Oslo, (2008)
 */
package de.Michael3600876;
```

```java
import java.util.*;

/*Implementation of IBAA rng by my brother*/
import de.alex4017769.*;
import java.io.*;

/**
 * @author Michael
 *
 */
public class Ising1D {
    /*Length of lattice should be power of 2*/
    private final int LENGTH = 100;

    /* determines initial spin configuration*/
    private boolean doCooling = false;

    private final int CYCLES = 1000;

    private int[] spins = new int[LENGTH];

    /* Temperature */
    private double T;

    /* External magnetic field */
    private double B;

    /* Energy of system */
    private double E = 0;

    /* E accumulator */
    private double EAcc;

    /* E^2 accumulator */
    private double E2Acc;
```

Implementation of the Ising models

```java
    /* Magnetization of system */
    private double M;

    /* M accumulator */
    private double MAcc;

    /* M^2 accumulator */
    private double M2Acc;

    /* Number of MC steps per spin */
    private int mcs = 0;

    /* # of accepted moves */
    private int accept;

    private Random r = new IBAA((int) System.
        currentTimeMillis());

/* 1d Ising lattice */


    public static void main(String[] args) {
        Ising1D test = new Ising1D();

        /* initial temperature */
        double x = 10;
        test.doInit(x, 0, true);

        double y = test.T;

        double stepsize = 0.01;

        /* Use this for additional precision */
        for (int i = 0; (y = x-i*stepsize) > 0; ++i) {
            test.setTemperature(y);
```

```
                test.metropolis();
                //test.printEnergy();
                //test.printMagn();
                //test.printSusc();
                test.printHeatCapacity();


                test.clearData();
        }}

        /**
         * Init functions that are called every time the temperature
           is changed.
         */

        /*
         * Initializes B, T and length of chain, calls function to
           create spins
         * calculates M
         */
                private void doInit(double T, double B, boolean
                    doCooling) {

                this.B = B;
                this.T = T;
                this.doCooling = doCooling;
                setExternalField(B);

                /* initializes array with spins */
                initSpin();

                /* calcluates respective magnetization */
                initMagn();

                /* calculates respective energy */
```

Implementation of the Ising models

```
            initEnergy();

            /* deletes accumulators */
            clearData();

        }


    /* Fills array with spins */
    private void initSpin() {
        for (int i = 0; i < spins.length; ++i) {
            if (true == doCooling){
                spins[i] = createRandomByte();
            }
            else {
                spins[i] = 1;
            }

        }
    }

    /* calculates magnetization */
    private void initMagn() {
        for (int i = 0; i < spins.length; i++)
        {
            M += spins[i];

        }

    }

    /* calculates internal energy with J = 1 */
    private void initEnergy () {
        int jm = spins.length - 1;
        for (int i = 0; i < spins.length; i++) {
            E -= spins[i] * spins[jm];
```

```
                    jm = i;
            }
            E -= M*B;
        }


        private void initDeltaE() {
        }

    /**
     * End of init functions.
     */


    /** Beginning of setter functions
     *
     */

    /**
     * set temperature
     */
        private void setTemperature(double T) {
            this.T = T;
        }

    /**
        * set B and change E accordingly
        */
        private void setExternalField(double B) {
            E += this.B*M - B*M;
            this.B = B;

        }

            /**
```

Implementation of the Ising models

```
         * set coupling constant j
         */

/*End of setter functions */

/*Function performing one metropolis step   */
    private void metropolis() {
        for (int k = 0; k < CYCLES; k++) {
            for (int i = 0; i < spins.length; ++i) {
            /*This is the Metropolis test */
             /* random pos on lattice */
                int pos = nextRandomIntOnLattice();
                double deltaE = calculateDeltaE(pos);
                if ((nextRandomDouble()<Math.exp(-deltaE/T) ||
                    deltaE <= 0)) {
                    doSpinFlip(pos);
                    accept++;
                    updateM(pos);
                    updateE(deltaE);

                }
            }

            /* update accumulators */
            mcs++;
            EAcc += E;
            E2Acc += E*E;
            MAcc += M;
            M2Acc += M*M;

        }

    }

/**
 * This function calculates the energy difference delta E
```

```
 * between two spin configurations.
 * @param pos
 * @return deltaE
 */
private double calculateDeltaE(int pos) {
    return 2 * (spins[pos] * (calculateNextNeigbhorSpinSum
        (pos)+B));
}
        /* resets all accumulators after one MC
                step per lattice site */
private void clearData() {
    accept = 0;
    EAcc = 0;
    E2Acc = 0;
    MAcc = 0;
    M2Acc = 0;
    mcs = 0;
}


/**
 * @param deltaE current deltaE
 */
private void updateE(double deltaE) {
    E += deltaE;
}

/**
 * @param pos current position on lattice
 */
private void updateM(int pos) {
    M += 2 *spins[pos];

}

/*
```

Implementation of the Ising models

```
 * Auxiliary functions
 */

/**
 * @param i
 * @return sum of next neigbor spins using
 * periodic boundary conditions
 */
    private int calculateNextNeigbhorSpinSum(int i) {
        return (spins[periodic(i - 1)] + spins[(periodic(i +
            1))]);

    }

/** This function flips the spin
 * @param currentSpin
 *
 */
    private void doSpinFlip(int pos) {
        spins[pos] = -spins[pos];

    }

/** This fucntion uses the IBAA to generate
 * rn
 * @return rn double
 */
    private double nextRandomDouble() {
        return r.nextDouble();
    }
/**
 * @return random byte -1 or 1 using
 * IBAA rng.
 * */
    private byte createRandomByte() {
        if (true == r.nextBoolean()) {
```

```java
                    return 1;
                }
                else
                    return -1;
            }
            /**
             * @return random int on the lattice
             * using IBAA rng.
             * */

                private int nextRandomIntOnLattice() {
                    return r.nextInt(spins.length);
                }

            /**
             * @param  value current position on lattice
             * without periodic boundary conditions.
             * @return position on lattice
             * enforcing periodic boundary conditions
             * in one dimension.
             */
                private int periodic(int value) {
                    return ((value + spins.length) % spins.length);
                }

            /*
             * End of auxiliary functions
             */

            /*Test function printing spins and their number */

                private void printSpins() {
                    for (int i = 0; i < spins.length; i++) {
                        System.out.println();
                        System.out.print("spin " + i + " has value: " +
                            spins[i]);
```

```
            }
            System.out.println("");
        }


        private void printMagn() {
            String sFileName = "mag1d.csv";

            try
            {
                File outFile = new File(sFileName);
                FileWriter writer = new FileWriter(outFile, true);
                String printString = "" + ((MAcc)/(spins.length*mcs
                    ));
                String temperature = "" + T;

                writer.append(printString + " ");
                writer.append(temperature);

                writer.append('\n');
                writer.flush();
                writer.close();
            }
             catch(IOException e)
            {
                e.printStackTrace();
            }

        }

        private void printSusc() {

            String sFileName = "susc1d.csv";

            try
```

```java
            {
                File outFile = new File(sFileName);
                FileWriter writer = new FileWriter(outFile, true);

                String printString = "" + (M2Acc-((MAcc*MAcc)/mcs))
                    /(T*mcs*spins.length);
                String temperature = "" + T;

                writer.append(printString + " ");
                writer.append(temperature);

                writer.append('\n');
                writer.flush();
                writer.close();
            }
            catch(IOException e)
            {
                e.printStackTrace();
            }

        }

        private void printHeatCapacity() {


            String sFileName = "theat1d.csv";

            try
            {
                File outFile = new File(sFileName);
                FileWriter writer = new FileWriter(outFile, true);

                String printString = "" + (E2Acc-(EAcc*EAcc)/mcs)/(
                    T*T*spins.length*mcs);
                String temperature = "" + T;
```

Implementation of the Ising models

```
            writer.append(printString + " ");
            writer.append(temperature);

            writer.append('\n');
            writer.flush();
            writer.close();
        }
         catch(IOException e)
        {
            e.printStackTrace();
        }
    }


    private void printEnergy() {

        String sFileName = "energy1d.csv";

        try
        {
            File outFile = new File(sFileName);
            FileWriter writer = new FileWriter(outFile, true);

            String printString = "" + (EAcc)/(spins.length*mcs)
                ;
            String temperature = "" + T;

            writer.append(printString + " ");
            writer.append(temperature);
            writer.append('\n');
            writer.flush();
            writer.close();
        }
         catch(IOException e)
```

```java
                {
                    e.printStackTrace();
                }
            }

            private void printAcceptRate() {

                String sFileName = "accept1d.csv";

                try
                {
                    File outFile = new File(sFileName);
                    FileWriter writer = new FileWriter(outFile, true);

                    String printString = "" + (double) ((accept)/(spins
                        .length*mcs));
                    String temperature = "" + T;
                    writer.append(printString + " ");
                    writer.append(temperature);

                    writer.append('\n');
                    writer.flush();
                    writer.close();
                } catch(IOException e)
                {
                    e.printStackTrace();
                }
            }

        }
```

B.3. Two dimensional Ising model

Listing B.2: Source code for 2D Ising model

```java
/**
```

Implementation of the Ising models

```
 * This is an implementation of the two-dimensional
 * Ising model without magnetic field using periodic
 * boundary conditions. Boltzmann's constant is set
 * one. Coupling constant can be set. => Temperature has thus
 * dimension energy. Metropolis sampling wih one site
 * spin flip is used. Part of the optimizations is described in
 * M. Hjorth-Jensen, Lecture Notes on Computational Physics,
 * University of Oslo, (2008)
 */

package de.Michael3600876;

import java.util.*;
import java.io.*;
/* Implementation of IBAA rng by my brother*/
import de.alex4017769.*;

/**
 * @author Michael
 *
 */
public class Ising2D {

    /*
    ****************************************************************
    */
    /**Beginning of variables that are user input and may be
       changed */

    /* Length of lattice should be power of 2 */
    private final int LENGTH = 100;

        /* determines initial spin config*/
    private boolean doCooling = true;

        /* Coupling constant */
```

```java
        private byte J = -1;

        /* # of cycles per MC step
         * without relaxation steps */
        private final int CYCLES = 10000;

/*relaxation time between measurements */
        private int relaxtime = 1;

/**End of variables that are user input and may be changed*/
/*************************************************************/

        /*plattform-dependent path separator*/
        private String pathSep = System.getProperty("path.
            separator");

        /*array containing spins */
        private int[][] spins = new int[LENGTH][LENGTH];

        /* Contains exp(-de/temp)
         * and magnetic field contribution */
        private double[][] weight = new double[17][3];
/*Observables*/
        /*******************************/
        /*** The following two observables
         * are parameters and are to
         * be set in the main function ***/

        /* Current temperature */
        private double T;

        /* External B field */
        private double B;

        /*******************************/
        /* Current Energy of system */
```

Implementation of the Ising models

```java
        private double E = 0;

        /* E accumulator */
        private double EAcc;

        /* E^2 accumulator */
        private double E2Acc;

        /* Current Magnetization of system */
        private double M;

        /* M accumulator */
        private double MAcc;


        /* M^2 accumulator */
        private double M2Acc;

        /* M^4 accumulator */
        private double M4Acc;

    /* Number of MC steps per spin */
    private int mcs = 0;

        /* # of accepted moves */
        private int accept;

    /*Determines whether relaxation steps are performed*/
    private boolean doRelax = false;

        private Random r = new IBAANative((int) System.
            currentTimeMillis());

    /* 2d Ising lattice (ferromagnet or antiferromagnet) */
```

```java
public static void main(String[] args) {
    Ising2D test = new Ising2D();
    /*Initial temperature saved in additional variable
    for higher precision*/
    double x = 2.6;
    double endTemp = 1;
    test.doInit(x, 5, true);
    test.initWeight();

    /*Current temperature*/
    double y = test.T;

    double stepsize = 0.5; //change if necessary


    /*use this for additional precision*/

    for (int i = 0; (y = x-i*stepsize) > endTemp; ++i) {

        test.setTemperature(y);
        test.initWeight();
        //if (y < 2.4) test.doRelax = true;
        test.metropolis();
        /*Uncomment functions to be performed*/
        //test.printBinder();
        //test.printEnergy();
        //test.printMagn();
        test.printSusc();
        test.printHeatCapacity();
        //test.printAcceptRate();

        test.clearData();

    }
```

Implementation of the Ising models

```
        }

  /**
   * Init functions that are called every time the temperature
     is changed.
   */

  /*
   * Initializes B, T and length of chain, calls function to
     create spins
   * calculates M
   */
        private void doInit(double T, double B, boolean
            doCooling) {

            this.B = B;
            this.T = T;
            this.doCooling = doCooling;
            setExternalField(B);
            initSpin(); // initializes array with spins
            initMagn();    // calcluates respective magnetization
            initEnergy(); // calculates respective energy
            clearData(); //deletes accumulators


        }


  /**
   * Function initializing weight[17]
   *    containing exp(-\beta\Delta{}E)
   **/
        private void initWeight() {
            for (int i = -8; i <= 8; i += 4) {
```

```
                    weight[i+8][0] = Math.exp((-i - 2*B)/T);  //TODO add
                        magnetic field term!!
                    weight[i+8][2] = Math.exp((-i + 2*B)/T);  //TODO add
                        magnetic field term!!

            }

        }



    /* Fills array with spins */
        private void initSpin() {
            for (int x = 0; x < spins.length; ++x) {
                for (int y = 0; y < spins.length; ++y) {
                    if (true == doCooling){
                        spins[x][y] = createRandomByte();
                    }
                    else {
                        spins[x][y] = 1;
                    }
                }

            }
        }

    /* calculates magnetization */
        private void initMagn() {
            for (int x = 0; x < spins.length; x++)
            {
                for (int y = 0; y < spins.length; y++) {
                    M += spins[x][y];
                }

            }
```

Implementation of the Ising models

```
        }

    /* calculates internal energy */
    private void initEnergy () {

        for (int x = 0; x < spins.length; x++) {
            for (int y = 0; y < spins.length; y++) {
                E -= .5*J*spins[x][y] *
                    calculateNextNeigbhorSpinSum(x, y) ;



            }
        }
        E -= M*B;
    }



    /**
     * End of init functions.
     */


    /** Beginning of setter functions
     *
     */

    /**
     * set temperature
     */
    private void setTemperature(double T) {
        this.T = T;
    }

    /**
     * set B and change E accordingly
```

```java
        */
        private void setExternalField(double B) {
            E += this.B*M - B*M;
            this.B = B;

        }


        /**
         * set coupling constant j
         */
        private void setCoupling(byte J) {
            this.J = J;
        }

    /*End of setter functions*/

    /*Function performing one metropolis step*/
        private void metropolis() {
          /*HACK: this implements relaxation steps */
            for (int j = 0; j < relaxtime; j++)
            {
            /*The last relaxation step is measured*/

                if (j == relaxtime - 1) {
                    doRelax = false;
                }

                for (int k = 0; k < CYCLES; k++) {
                    for (int i = 0; i < (spins.length * spins.length
                    ); ++i) {
                    /*This is the Metropolis test*/
                        int posx = nextRandomIntOnLattice();// random
                            pos on lattice
                        int posy = nextRandomIntOnLattice();// random
                            pos on lattice
```

```
                    double deltaE = calculateDeltaE(posx, posy);
                    if ((nextRandomDouble() <( weight[(int)
                        deltaE + 8][1 + spins[posx][posy]])
                    ||(deltaE - 2*spins[posx][posy]*B <= 0)))
                    {
                        doSpinFlip(posx, posy);
                        //accept++;
                        if (false == doRelax) {
                            updateM(posx, posy);
                            updateE(deltaE);
                        }


                    }
                }

                /* update accumulators */
                if (false == doRelax)
                {
                    mcs++;
                    EAcc += E;
                    E2Acc += E*E;
                    MAcc += M;
                    M2Acc += M*M;
                    //M4Acc += M*M*M*M;


                }
            }

        }



    }
```

```
/**
 * This function calculates the energy difference delta E
 * between two spin configurations.
 * @param pos
 * @return deltaE
 */
private double calculateDeltaE(int x, int y) {
    return 2 *J* (spins[x][y] * (
        calculateNextNeigbhorSpinSum(x, y)));
    //return (2 * spins[pos]*(B+spins[(pos+1)%spins.length
    //  ] + spins[(pos-1+spins.length)%spins.length]));
}

/* resets all accumulators */
private void clearData() {
    //accept = 0;
    EAcc = 0;
    E2Acc = 0;
    MAcc = 0;
    M2Acc = 0;
    //M4Acc = 0;
    mcs = 0;

}
```

Implementation of the Ising models

```java
/**
 * @param deltaE current deltaE
 */
private void updateE(double deltaE) {
    E += deltaE;
}

/**
 * @param pos current position on lattice
 */
private void updateM(int x, int y) {
    M += 2 * spins[x][y];

}

/*
 * Auxiliary functions
 */

/**
 * @param i
 * @return sum of next neigbor spins using periodic boundary
 *   conditions
 */
private int calculateNextNeigbhorSpinSum(int x, int y) {
    return (spins[periodic(x - 1)][y] + spins[(periodic(x
        + 1))][y]
        + spins[x][periodic(y - 1)] + spins[x][(periodic(y
        + 1))]);

}

/**This function flips the spin
 * @param currentSpin
 *
```

```java
     */
    private void doSpinFlip(int x, int y) {
        spins[x][y] = -spins[x][y];

    }

    /**This fucntion uses the IBAA to generate
     * rn
     * @return rn double
     */
    private double nextRandomDouble() {
        return r.nextDouble();
    }
    /**
     * @return random byte -1 or 1 using
     * IBAA rng.
     * */
    private byte createRandomByte() {
        if (true == r.nextBoolean()) {
            return 1;
        }
        else
            return -1;
    }
    /**
     * @return random int on the lattice
     * using IBAA rng.
     * */

    private int nextRandomIntOnLattice() {
        return r.nextInt(spins.length);
    }

    /**
     * @param value current position on lattice
     * without periodic boundary conditions.
```

```
 * @return position on lattice
 * enforcing periodic boundary conditions
 * in one dimension.
 */
    private int periodic(int value) {
        return ((value + spins.length) % spins.length);
    }

/*
 * End of auxiliary functions
 */

/* Test function printing spins
 * and their number */

    private void printSpins() {
        for (int x = 0; x < spins.length; x++) {
            for (int y = 0; y < spins.length; y++) {
                System.out.println();
                System.out.print("spin ("  + x + ", " + y + ") "
                    + " has value: " + spins[x][y]);


            }
        }
        System.out.println("");
    }

    /* Function printing spins as they are on the lattice
     */

    private void printSpinConf() {
        for (int x = 0; x < spins.length; x++) {
            for (int y = 0; y < spins.length; y++) {
                System.out.print(spins[x][y]+" ");
```

```java
            }
            System.out.println("");
        }
    }


    private void printMagn() {
        String sFileName = "testmag.csv";

        try
        {
            File outFile = new File(sFileName);
            FileWriter writer = new FileWriter(outFile, true);
            String printString = "" + ((Math.abs(MAcc)))/(mcs*
                spins.length*spins.length);
            String temperature = "" + T;

            writer.append(printString + " ");
            writer.append(temperature);
            writer.append('\n');
            writer.flush();
            writer.close();
        }
          catch(IOException e)
          {
              e.printStackTrace();
          }


    }

    private void printSusc() {

        String sFileName = "testsusc.csv";
```

Implementation of the Ising models

```java
        try
        {
            File outFile = new File(sFileName);
            FileWriter writer = new FileWriter(outFile, true);

            String printString = "" + (M2Acc-((MAcc*MAcc)/mcs))
                /(T*mcs*spins.length*spins.length) ;
            String temperature = "" + T;

            writer.append(printString + " ");
            writer.append(temperature);
            writer.append('\n');
            writer.flush();
            writer.close();
        }
         catch(IOException e)
        {
             e.printStackTrace();
        }


    }

    private void printHeatCapacity() {

        String sFileName = "testheat.csv";

        try
        {
            File outFile = new File(sFileName);
            FileWriter writer = new FileWriter(outFile, true);

            String printString = "" + (E2Acc-(EAcc*EAcc)/mcs)/(
                T*T*spins.length*spins.length*mcs);
            String temperature = "" + T;
```

```
                writer.append(printString + "␣");
                writer.append(temperature);
                writer.append('\n');
                writer.flush();
                writer.close();
            }
            catch(IOException e)
            {
                e.printStackTrace();
            }
        }


        private void printEnergy() {

            String sFileName = "testen.csv";

            try
            {
                File outFile = new File(sFileName);
                FileWriter writer = new FileWriter(outFile, true);

                String printString = "" + (EAcc)/(spins.length*
                    spins.length*mcs);
                String temperature = "" + T;

                writer.append(printString + "␣");
                writer.append(temperature);
                writer.append('\n');
                writer.flush();
                writer.close();
            }
            catch(IOException e)
            {
```

```java
                    e.printStackTrace();
                }
        }

        private void printAcceptRate() {

            String sFileName = "test2.csv";

            try
            {
                File outFile = new File(sFileName);
                FileWriter writer = new FileWriter(outFile, true);

                String printString = "" + (double) ((accept)/(spins
                    .length*spins.length*mcs));
                String temperature = "" + T;

                writer.append(printString + " ");
                writer.append(temperature);
                writer.append('\n');
                writer.flush();
                writer.close();
            }
             catch(IOException e)
             {
                 e.printStackTrace();
             }
        }

        private void printBinder() {

            String sFileName = "bindercumul2.csv";

            try
            {
                File outFile = new File(sFileName);
```

```
            FileWriter writer = new FileWriter(outFile, true);
            String printString = "" + (-(mcs*M4Acc)/(3*M2Acc*
                M2Acc)+1) ;
            String temperature = "" + T;

            writer.append(temperature + "␣");
            writer.append(printString);
            writer.append('\n');
            writer.flush();
            writer.close();
        }
         catch(IOException e)
         {
             e.printStackTrace();
         }

    }

}
```

Bibliography

1. D. Arovas, *Lecture Notes on Thermodynamics and Statistical Mechanics (A Work in Progress)* (University of California San Diego, 2010).
2. N. W. Ashcroft, N. D. Mermin, *Solid State Physics* (Brooks Cole, 1976).
3. S. Backes, Private communication.
4. R. J. Baxter, *Exactly Solved Models in Statistical Mechanics* (Dover Publications, 2008).
5. K. Binder, *Zeitschrift für Physik B Condensed Matter* **43**, 119–140 (1981).
6. N. Blümer, *Moderne numerische Methoden der Festkörperphysik (lecture notes)* (Universität Mainz, 2008).
7. H. Gould, J. Tobochnik, W. Christian, *An Introduction to Computer Simulation Methods: Applications to Physical Systems (3rd Edition)* (Addison Wesley, 2006).
8. W. K. Hastings, *Biometrika* **57**, 97–109 (1970).
9. H. Haug, *Statistische Physik: Gleichgewichtstheorie und Kinetik* (Springer, 2005).
10. M. Hjorth-Jensen, *Computational Physics (lecture notes)* (University of Oslo, 2009).
11. E. Ising, *Zeitschrift für Physiker* **31**, 253–258 (1925).
12. H. G. Katzgraber, *Inroduction to Monte Carlo Methods* (ETH Zürich, 2009).
13. D. P. Landau, K. Binder, *A Guide to Monte Carlo Simulations in Statistical Physics* (Cambridge University Press, 2009).
14. W. Lenz, *Physikalische Zeitschrift* **21**, 613–615 (1920).
15. D. Meier, *Bachelorarbeit: Untersuchung des zweidimensionalen Ising-Spinsystems auf Phasenübergänge* (Universität Bielefeld, 2009).
16. N. Metropolis, A. W. Rosenbluth, M. N. Rosenbluth, A. H. Teller, E. Teller, *The Journal of Chemical Physics* **21**, 1087–1092 (1953).

17. M. E. J. Newman, G. T. Barkema, *Monte Carlo Methods in Statistical Physics* (Oxford University Press, USA, 1999).

18. W. Nolting, *Grundkurs Theoretische Physik 6: Statistische Physik* (Springer, 2005).

19. L. Onsager, *Phys. Rev.* **65**, 117–149 (1944).

20. T. Pang, *An Introduction to Computational Physics* (Cambridge University Press, 1997).

21. V. Privman, Ed., *Finite Size Scaling and Numerical Simulation of Statistical Systems* (World Scientific Publishing Company, 1990).

22. F. Reif, *Statistische Physik und Theorie der Wärme* (Walter de Gruyter, 1987).

23. A. Sokal, *Monte Carlo Methods in Statistical Mechanics: Foundations and New Algorithms (lecture notes)* (University of North Carolina, 1996).

24. U. Wolff, *Phys. Rev. Lett.* **62**, 361–364 (1989).

List of Figures

3.1. Magnetization 1D Ising model theoretical and MC results 30

3.2. Magnetic susceptibility for $B = 0$ and $B = 1$ theoretical and MC results . 31

3.3. Energy 1D Ising model theoretical and MC results 32

3.4. Heat capacity for $B = 0$, $1\,2$ and $B = 10$ from MC simulation and exact result without fiel

3.5. E and M as a function of MC steps with initial configuration $T = \infty$ . . 33

3.6. E in two different initial spin configurations as a function of MC steps . . 34

3.7. M in two different initial spin configurations as a function of MC steps . . 34

3.8. Magnetization for different lattice sizes 36

3.9. Energy for different lattice sizes 36

3.10. Magnetic susceptibility for different lattice sizes 37

3.11. Heat capacity for different lattice sizes 37

3.12. Energy of 2D Ising ferromagnet for different B 38

3.13. Magnetization of 2D Ising ferromagnet for different B 39

3.14. Susceptibility of 2D Ising ferromagnet for different B 40

3.15. Heat capacity of 2D Ising ferromagnet for different B 41

3.16. Magnetization of 2D Ising antiferromagnet for different B 42

3.17. Magnetization of 2D Ising antiferromagnet for $B = 0$ 42

3.18. Energy of 2D Ising antiferromagnet for different B 43

3.19. Susceptibility of 2D Ising antiferromagnet for different B 43

3.20. Heat capacityof 2D Ising antiferromagnet for different B 44

3.21. Fourth order Binder cumulant for different L 46

3.22. Fourth order Binder cumulant against size–scaled reduced temperature . 47

B.1. Illustration of periodic boundary conditions (*17*) 56